The Answer

THE ANSWER

WHY ONLY INHERENTLY SAFE, MINI NUCLEAR POWER PLANTS CAN SAVE OUR WORLD

Reese Palley

The Quantuck Lane Press
NEW YORK

Layout and Composition:
   John Bernstein Design, Inc.
Manufacturing by
   Maple-Vail Manufacturing Group
All rights reserved
Printed in the United States
First Edition

Library of Congress
   Cataloging-in-Publication Data

Palley, Reese.
The answer : why only inherently safe,
mini nuclear power plants can save our
world / Reese Palley.

   p.   cm.
   ISBN 978-1-59372-045-2
1. Nuclear energy — United States.
2. Nuclear power plants — Safety measures
— United States. 3. Global warming —
United States.   I. Title.
   TK9146.P35 2011
   333.792'4 — dc22

                              2011008501

The Quantuck Lane Press, New York
www.quantucklanepress.com

Distributed by:
W.W. Norton & Company
500 Fifth Avenue, New York, NY 10110
www.wwnorton.com

W. W. Norton & Company Ltd.
Castle House, 75/76 Wells Street
London, WIT 3QT

1 2 3 4 5 6 7 8 9 0

OTHER BOOKS BY REESE PALLEY

Concrete: A Seven-Thousand-Year History by Reese Palley, The Quantuck Lane Press, 2010

Wooden Ships and Iron Men: The Maritime Art of Thomas Hoyne by Reese Palley and Marilyn Arnold Palley, The Quantuck Lane Press, 2005

The Best of Nautical Quarterly, Vol. I: The Lure of Sail by Reese Palley and Anthony Dalton, MBI International, 2004

Call of The Ancient Mariner, McGraw Hill / International Marine, 2004

Unlikely People, Sheridan Press, 1998

There Be No Dragons, Sheridan Press, 1996

Unlikely Passages, Simon & Shuster / Seven Seas Press, 1984

The Porcelain Art of Edward Marshall Boehm, Harry N. Abrams, 1976

Contents

Dedication

DOCTORS Gluckman, Ferrari, and Kochman, with an assist from Scissorhands Acker, HUP men all, are the team that kept me alive long enough to finish this book. It was not only their medical skills that did the deed but their very human interest and support for the job of work that the book represents. At one point, when age and ennui suggested that the burden of finishing the book was too great, Vic Ferrari, a gentle man not comfortable with ordering folk about, set me sharply back on track.

Acknowledgments

IT IS A PLEASURE to acknowledge a few of the folks who gave that extra push when the last chapter seemed impossible years away. First I need to pay homage to my publisher, Jim Mairs, who, luckily for me, after a lifetime career at Norton decided to strike out on his own and establish The Quantuck Lane Press, the publisher of my last two books. He is followed by Austin O'Driscoll who picked up many of the pieces and John Bernstein the designer who put the pieces together. Then there is Bob Sadlowe who led me through so many nuclear mazes as the book began to take shape and Jeff Carpenter who gave the raw manuscript the closest read that could be imagined and saved me embarrassment and dismay and Tony Auth who kept me laughing and partially sane in our troubled world.

Of course there is my wife Marilyn....there is always my wife Marilyn, who danced about me with the energy of a small nuclear device, encouraging, correcting, admonishing and, best of all, loving me into finishing the book.

There were others, too many to mention, who added their mite. But, in the end, it was the Random Universe which drenched me with serendipity and granted me all the lucky accidents without which little, let alone this book, could ever get done.

Foreword

THIS BOOK WAS conceived and researched three years before the sea rose up and smote the northeast coast of Japan and the now-infamous reactors at Fukushima.

The odds are very long that the rest of the world will extrapolate from Japan's ill luck a long-term plan to deal with the warming of the Earth. Such a long-term plan would be the best possible outcome of the Japanese tragedy. Equally heartening would be a saner, less danger-filled approach to the use of nuclear power along the lines outlined in this book. Whether the motivation is the saving of the planet, or just a major step forward in the safe generation of inexpensive, critically needed electrical power, this book offers a practical, reasonable alternative to the choices we have made, and continue to make around nuclear power generation.

Worse than simply doing nothing is a scenario in which the Fukushima meltdowns lead us to turn away from nuclear energy altogether and force us back to burning even more and more fossil fuels. From the initial public reactions throughout the U.S. and Europe, this may very likely be the outcome.

If we do not, with all haste, solve the safety concerns of the nuclear industry and, at the same time, call a halt to the enormous flow of CO_2 into our burdened, atmosphere, there will be no avoiding a dismal future for man on earth. By continuing to build vast, inherently unsafe reactors we are acting out Albert Einstein's definition of insanity: repeating the same behavior while expecting a different outcome.

This book offers a safe and non-abusive approach to our burgeoning hunger for energy and posits a reasonably predictable and positive outcome. What we lack is not awareness of the enormity of our peril, merely the will to address it.

Argument

THE intent of this book is to make clear, in lay terms, that our future on Earth is inextricably embedded in a particular profile of nuclear energy. It will be further argued that the opposition to nuclear energy derives from problems of scale and safety, problems that are within our present technological capabilities to resolve. It will also be argued that financial risk, exacerbated by the unknowns inherent in huge-scale sequestration of capital over long periods of time, is a chief inhibitor in the growth of sufficient nuclear energy needed to combat the worst eventualities of global warming.

The nuclear era was thrust upon us only in the past seventy years. So short a time frame is out of sync with the geologic time frame in which the history of the Earth is recorded. Seventy years has no impact on the natural processes that regulate how the planet accommodates change. Geologically we deal in billions of years, while the time frame that defines all living things is mere millions and, in the case of modern man, some few thousands of years. Neither Earth nor the biota that inhabit it has any built-in, self-regulating mechanisms for dealing with a fundamental planetary change that occurs in the space of only one generation of mankind, a paltry speck of time.

Planetary events are glacially slow. The ice age, ringing its changes in thousands of years, is, in the context of the developing Earth, as sudden as a lightning strike. These cold centuries, by their nearness in history, are still very much with us in our calculations of weather, in our definitions of human comfort, and in our understanding of the geology of our planet. They lowered the sea level by scores of feet, killed off thousands of species, and left our own upright big-brained ancestors with only a few thousand breeding pairs. It is this relationship between solid and liquid water, so important in climate, that is changing with worrisome speed.

When geologic changes take millions or billions of years, there is plenty of time for living forms to accommodate. Lewis Thomas, in *Lives of a Cell*, viewed our planet as a single-celled animal in constant adjustment to changing internal forces. Barring external events, and the time to adjust to internal events, our planetary home would roll happily along, balancing out the gentle conflicting claims and requirements of its living inhabitants.

Indeed, our planetary history has progressed more or less quietly, give or take a few meteors, for some billions of years. Checks and balances kept things on a steady path until, in relatively recent times, a tool-using avatar with an opposable thumb appeared and slowly started to bend the environment to suit his needs. Man on Earth was an acceptable condition for the planet until, aided by an expanding brain and the discovery of agriculture, he went viral about three hundred years ago and, like all viruses, commenced an attack on his host.

The planet had accommodated itself to man for some hundreds of thousands of years, during which the changes induced by the human virus were slow. The problem was not change in itself. The problem was the rate of change. When human population suddenly went viral around 1800 AD, we altered the ancient relationship in which man accommodated to settled natural rhythms. Now, due to a burgeoning population and the inventions of the industrial revolution, the alteration of the *rate of change* in the relation between carbon sequestered and carbon emitted is straining, in deadly fashion, the planet's ability to accommodate to the ever accelerating activities of industrial man. The virus is killing its host, and if the host dies too quickly man dies with it. The real problem was, and remains, the *rate of change*.

The resources of the planet, unassisted, easily supported three hundred million humans, which was its population preindustrialization and when relatively little acreage had been cleared for agriculture. But when population leapt in a few ticks to six billion, support systems began to break down, and the effluvia of our industrial inventions, along with changes in our eating habits, mounted the viral attack on our earthly host.

It can be argued that, if the growth in population from fewer than three hundred million to six billion had occurred over a period of 200,000 instead of 2,000 years, all sorts of balancing processes would have developed. These would have involved weather, food sources, energy generation, control of effluvia, and new technologies. In light of limited resources, such processes could have helped establish and maintain an acceptable mutuality and indefinitely continuing Earth's use as a habitat for humanity.

Global warming, traceable to one particular class of effluvia of human activities, greenhouse gases (GHGs), became a matter of serious concern in only the past few decades. Predicted and proven rises in ambient world temperatures are now making a hard promise that man's future on Earth will be so different from his past as to be unrecognizable. Global warming, created by our teeming and ever rising billions, is a threat to our existence within the time frame in which we and our children will be living out our lives. We big brains, and the rest of the myriad life with which we share the planet, live as we shall see within a remarkably narrow set of physical parameters only one of which is temperature but all of which are related to heat.

Fifty years ago, serendipitously and surprisingly, nuclear sources for power generation emerged that can provide a previously nonexistent antidote to global warming. The explanation of how two such antipathetical events, global warming and an antidote, should appear at roughly the same moment in history is a teleological question best left to the lords of the church or to the lords of chaos. In any event, it is vastly beyond the scope of this book.

Whoever or whatever is responsible for handing us the problem of global warming and the solution of nuclear power on the same platter of history—it is a coincidence of staggering proportions. Had we been a soupçon out of sync, even by a handful of years, we would certainly have continued down the path of catastrophically carbonizing our little blue orb.

While warming, arising from man's burning of carbon, suggests the end of human society as we know it, the parallel appearance of nuclear energy, independent of that central activity of all living things, the burning of carbon, is either a miracle or one hell of an accident.

This improbable historical mystery has handed us a lifeline to a future that might be not too unlike our comfortable past. However, if we fail to grasp the line, if we fail, in the very short time that is left to us, to correct the incendiary naïveté of our past, our future will be at best very uncomfortable and at worst simply unlivable.

After World War II, when Robert Oppenheimer was asked if he believed that humankind could survive a nuclear war, he answered, "That depends on how you define what is human."

Now, when the same question is asked—whether humankind can survive the looming planetary response to our past profligate foolishness—the same answer will serve.

Introduction

DURING THE WAR YEARS prior to 1945 some thousands of scientists, newly minted as *nuclear* scientists, scurried around under the darkest of secrecy blankets to convert white chalk lines on a black board into a living, pulsing ball of energy that would change the world.

In 1945, like Magi bearing gifts, they deposited, at the feet of a handful of generals and politicians, a barely proven bomb of boundless destructive power. It ultimately landed with a thud on the desk of the buck-stops-here president.

During the heady and passionate work of transmuting theory into fact, the moral considerations of making a weapon with the seeds of mass annihilation were deferred to the future. The insistent curiosity of scientists, the fear of Nazi acquisition, and the immediate pressure of the needs of the military inhibited consideration of the implications of developing world-eating weapons. Threads of disquiet disturbed some of the brainy craftsmen who, like Oppenheimer, in the rush of war, had reserved moral judgment but who now were faced with enabling an instant mass killing on a scale not heretofore imagined. Judgment could no longer be reserved.

The moral question that blossomed in 1945 was whether it was necessary to drop the bomb or whether the simple knowledge of the existence of such a weapon would be enough to bring about Japanese surrender. The experience of the Pacific war, in which hundreds of thousands of Japanese troops hopelessly gave their lives in defense of mere specks of useless atolls, suggested that an attack on the Japanese homeland would be met by fierce, determined, and, what is worse, irrational resistance.

President Truman was told that an invasion of Japan would cost a million American lives and some millions more wounded and would, after subjugating the Japanese military, require a long, debilitating occupation that might well generate anti-American terrorism. He was also told that if an essentially untested atomic bomb was dropped on a deserted atoll as a demonstration to the Japanese and it did not perform, the war with Japan would be extended by years.

On the other hand, if the bomb was deployed on a Japanese city, the cost would be in the hundreds of thousands of "less precious" enemy lives. Truman dropped the bomb and blew a hole in history.

The Hiroshima bomb was a device that had the explosive power of a few thousand tons of TNT. Within a few years the relatively meager output of the early atomic bombs was blindingly eclipsed by a weapon that owes much of its origin and development to one man.[1] Edward Teller arrived from Hungary at the moment when the United States was gearing up for its cold war conflict with the USSR. It was Edward Teller who, in 1939, supported his fellow Hungarian Leo Szilard when Szilard alerted Albert Einstein to the possibility that Hitler was working on an atomic weapon. Teller, Szilard, and Einstein, all Jews, were motivated by reports leaking out of Nazi Germany concerning an unthinkable extermination of European Jewry. The resulting historical letter of August 2, 1939, from Einstein to President

Roosevelt lit the fuse. With Roosevelt's backing, the Manhattan Project blossomed into a period of intense, rewarding, and exhilarating scientific research. The engine of this research was Einstein's original suggestion that there was a relationship between a small amount of matter and an enormous amount of energy. His immortal formula $E = mc^2$, where the multiplier c^2 is the speed of light, m is mass, and E is energy, had set physicists around the world in search of a way to winkle out this enormous power of the atom.

These first bombs, as measured by power released, were soon left in the dust by new work by Teller which led directly to a weapon many orders of magnitude beyond the early bomb, variously called a thermonuclear or hydrogen bomb. Thermonuclear bombs, ignited by old-fashioned fission bombs, had no theoretical limit in either size or destructiveness. Teller's hydrogen bomb could be built to release the power of however many millions of TNT that one would require.

The race for destructive capacity was on. The USSR and the United States, in a mad dash for military superiority, quickly accumulated tens of thousands of hydrogen bombs, each of which was thousands of times more powerful than the atomic bomb that destroyed Hiroshima.

Our largest thermonuclear bomb was "Ivy Mike," detonated in the Pacific in 1952. The bomb went off on the island of Elugelab and produced the equivalent of over ten million tons of TNT. When the smoke cleared, observers were startled to find that the entire island had simply disappeared. Elugelab, some millions of tons of earth and rock, had been lofted as dust into the air and wafted across the Pacific.

Plutonium, the fuel of the hydrogen bomb, was at that time rare and expensive to produce. When it became clear that there were enormous economies of scale in the production of plutonium, the course of the next half century of nuclear power was set in radioactive stone. This early military need for plutonium led to huge-scale breeder reactors, which produced more radioactive material than they consumed. It was these breeder reactors, the progenitors of almost all currently operating nuclear power reactors, that created the technology of size that dominates current thinking concerning the production of electrical power from atoms.

Historically, technologies that get an early start in an industry quickly come to dominate the development of that industry and as quickly tend to shut out promising new or parallel technologies. After some millions of engineering hours and some billions of private and public dollars have been invested, it would take a draconian upheaval in the way America does business to alter the course of the presently embedded and invested nuclear technology.

What further diminishes the possibility of real change in basic nuclear technologies in the United States today is the fanatic regulatory mechanism that grew out of the nuclear pandemics ignited by Three Mile Island and Chernobyl. Consequent to these events, regulatory approval of even a plant using a well-known technology already in operation required decades and billions of dollars to achieve.

Thus everything is stacked in favor of the existing large scale image of the production of energy from the atom. That profile is defined by the beneficial economies of size that are enjoyed by the private producers of nuclear energy and ignores the public costs of the entrenched technology.

The central argument of this book will be to demonstrate that the deleterious effects of existing huge 1,000-megawatt light water reactors (LWRs), of which there are hundreds in the planning stage, can be eliminated by the use of small, modular 100-megawatt nuclear power plants. Many of these mini–nuclear plants, SMRs (for small and medium-sized reactors), have been in successful and safe, and limited, operation for almost fifty years. This small-unit technology, outlined by Teller around 1950, is only beginning to be conjured with on an industrial scale and still has mountains of regulatory heights to climb.

Global warming is a megaton economic bomb that has renewed the drive to replace coal and oil with nuclear power. But, as we have seen, the thrust of the industry has been in the direction of the existing technologies of large-scale breeder reactors. These reactors reflect the known economic risks that a private capital market requires, as well as a guarantee of final regulatory approval. However, these known economic risks to capital have been recently called into question by the accelerating increase in the costs and the time required to build a megawatt reactor. With private capital chary of anything new or unknown in the nuclear industry and a regulatory system that, by its own admission, is neither equipped nor inclined to put standards into place for small reactors, huge hurdles remain that resist our rethinking of quick and cheap sources of nuclear energy. Thinking is hard; rethinking, in a field fraught with emotional responses, technical jargon, and protection of careers, is from daunting to impossible.

Complicating everything are the political and geo-political considerations represented by the demands of emerging nations hungry for energy from any source. The fastest way for these countries who have a less strident regulatory structure to get electricity online is to buy existing gigawatt technology, which is exactly what they are doing.

This book will demonstrate that the application of a slight fraction of the finances and intellectual work being expended on huge nuclear power plants to small, inherently safe, factory-built, truck-deliverable, modular mini–nuclear power plants will create solutions to the intractable and unintended consequences that accompany all megawatt-scale nuclear generation.

This book has two goals. The first is to argue that all nonnuclear sources of electric power either directly or indirectly produce carbon dioxide (CO_2) or directly or indirectly have warming consequences that are unacceptable.

The second goal is to argue that the nuclear power generation upon which we depend is, due to its inhuman scale, too awkward, too expensive, too long in coming, and too scary.

Because of time and money constraints, if we are to contain the threat of too much CO_2 in our fragile atmosphere, we must reduce the size of our nuclear plants to the scale and the ease of a backyard industry.

CHAPTER ONE

The Burning Question

"Global warming could be one of humankind's longest lasting legacies. The climatic impacts of releasing fossil fuel CO_2 to the atmosphere will last longer than Stonehenge. Longer than time capsules, longer than nuclear waste, far longer than the age of human civilization so far. Each ton of coal leaves CO_2 gas in the atmosphere. The CO_2 coming from a quarter of that ton will still be affecting the climate one thousand years from now, at the start of the next millennium. And that is only the beginning."

—David Archer, *The Long Thaw*

The universal formula that defines animal life on Earth is: Carbon + Oxygen = Heat + CO_2

IT IS NOT money that makes the world go round, it is this seemingly innocent little formula. *Everything* that has any spark of animal life in it depends on its ability to produce or use CO_2.

Humans, animals, fish, plants, and bacteria are all factories that take in oxygen, do a little magic, extract a little energy, and expel that devilish molecule carbon dioxide out of our variously and curiously formed smokestacks. The little energy that is extracted in the process is another word for *life*. To have life, as it has been defined on our planet, there is no escaping the use of CO_2.

We, the entire biota of our planet, take one atom of carbon and two atoms of oxygen and create an effluvium that will, it is guaranteed, raise the temperature of Earth and thrust our planet home back to an age when there was no ice. Without ice everything changes, but more of that later.

The cycle by which CO_2 is emitted into our atmosphere has been millennially balanced by life processes and planetary entities that take in and sequester the gas. Trees, for example, are a great eater of CO_2 and hold it tight until, by either burning or rotting, it escapes back into the air. The oceans, a major player, breathe in and out and hold on to vast quantities.

In the geological history of our planet, a balance between production and sequestration of CO_2 created a set of conditions that led to the explosion of the plentiful life all around us. Wherever there is a niche, however small, for the opportunity to mix carbon with oxygen, life has arisen. For all of the history of life on Earth there has been a nice mechanism that kept earthly temperatures approximately tuned to the requirements of the species that share the planet. When conditions changed—when there were incremental changes in temperature or salinity or cloud cover—innumerable species died and countless others arose. But, on balance, and assuming that none of the dominant life-forms were too smart, the Earth remained a fecund, balanced, and hospitable nest for carbon-based life.

Everything changed when blind evolution made the game-changing error of favoring sentient brains over unconscious brawn. Brawn can establish hegemony over fellow species, but brains, as it is turning out, were capable of establishing hegemony over the Earth itself.

We, the brainy ones, became the major players in creating an unimaginably long unintended tail of disasters into the future. Our grandchildren will begin to pay heavily for our gluttony for carbon. Ten thousand or a hundred thousand years from now humankind, if we are still around, will still be paying the bill for that as long as CO_2 can remain in the atmosphere.

So long as the source of carbon came from current and balanced and limited sources such as the effluvia of living things and the rotting of biotic material and an occasional volcanic eruption, the process contained no threat to existence. Carbon supplies normally available for burning and sequestration and reuse created an essentially renewable cycle. Nature had developed an environmentally acceptable way for life to be lived, and, until only a few nano moments ago in the life of the planet, uncounted natural processes of emission and sequestration of carbon danced about in perfect array.

Then we really smart humans discovered a secret inheritance. Carbon-bearing molecules that had lived and died for hundreds of millions of years were buried within our reach. Immediately there was an explosion in the availability of carbon for burning. It took only a brief time for man, assisted by drag lines as big as the Ritz, drilling towers as ubiquitous as trees, and mines as deep as hell, to haul out this sequestered carbon, coal, oil, and natural gas and put it to work on a grand scale.

Any of the problems that result from burning are essentially a matter of scale. The natural balance of emission and sequestration had some small ability to manage an increase in carbon dioxide over millennial periods of time. But the death of the trillions of life-forms that created coal and its cousins had, over thousands of millennia, buried carbon on a scale that proved too hot for nature to handle.

The conversion of this newly discovered high-energy source by burning to create steam and then a few years later to make electrical energy, began a two century-long growth of CO_2 emissions for which no natural reciprocating sequestration mechanism was available.

For two hundred years, new industrial processes and a burgeoning population introduced greenhouse gases that had never been significant in the natural release and recapture cycle for the life of the Earth. The most egregious of the gases is CO_2, because of the billions of tons involved. But there are others: methane and nitrous oxide, as well as the fluorocarbons and halogens, which, while churned out in lesser quantities, are proving in some cases to be more insidious in their atmospheric effects than CO_2.

In addition to the CO_2, we have let loose this host of abusive gases and particles that enormously complicate any attempt to backpedal from a possible tipping point. All must be dealt with and, further complicating matters, each requires a specific set of solutions. The fight against global warming is not merely a battle. It is much worse. It is a potential massacre.

There are fundamental problems involving the change in the makeup of our atmosphere that must be understood, especially by the nonscientific community, in order for us to marshal effective opposition to global warming.

Some of the methane released from landfills is being captured and used for energy production. This reduces the amount of methane, but it increases the amount of CO_2 (resulting from the burning of the captured methane). While CO_2 adds less to warming than does methane, it lasts much longer in the atmosphere.

In addition to the release of methane from livestock manure, the flatulence of billions of cattle fed on gas-producing feed is a dangerous source of methane. (Humans, too, produce methane-rich flatulence.) Fluoride gases such as the halogens, long known for the destruction of the ozone layer, are spectacular contributors to warming.[2] Fluorides have a short life of thirty-six years, but during that time they are almost 5,000 times more damaging to the atmosphere than CO_2.

The concrete industry is high up among the producers of CO_2. A few centuries ago we used very little concrete. Our consumption of this ubiquitous material now is gargantuan and growing. Some observers claim that the concrete industry's production of CO_2 has risen to 18 percent of the 8 billion tons of carbon that we loft annually. Cement, the essential constituent of concrete, is made by burning limestone with fossil fuels.

Every industrial, agricultural, and extraction industry in our modern economy is a producer of CO_2, either directly or indirectly. Based as these industries are on the use of biofuels, there is no tweaking or capping that is likely to have much effect on our warming problem. The complexity of our tightly interrelated economy suggests that only *complete abstention* from the burning of fossil fuels could prevent our planet from evolving into a mild sort of hell—and even abstention is not a guaranteed solution.

3) Quantity of Greenhouse Gases

There is one simple fact defining the level of greenhouse gases in the atmosphere and it deals with events that have not yet happened. That fact is that the population in generations to come will increase exponentially. We have established that GHGs are due essentially to the intervention of man into the eons-long cycles between the planet and its biosphere. We are now six billion people, churning out CO_2 via our very clever machines, with its production further exacerbated by our eating of meat animals. We are an obedient race and take seriously the biblical exhortation to go forth and multiply. There seems to be no natural mechanism capable of halting the growth of six billion to the predicted nine billion in a single generation. By 2050 we will have added 50 percent to the world's population, and there is no question but that this additional population will make at least half again as much CO_2 as we are making now.

Considering population growth, if the present looks bad, the future—the very, very near future—looks dreadful.

Since around 1800, the last "natural" year in which planetary CO_2 emissions were not seriously assisted by mankind, CO_2 has grown to 8 billion tons a year. If this level of emission were to continue unabated, it will drive the present 388 parts per million (ppm)[3] of CO_2 to a tipping point of 450 ppm, beyond which we would have little chance to develop methods to reverse the process of global warming.

There is some disagreement on what the actual release tonnages are. Much depends on what sources of energy are included and which are excluded. There are also no fixed yardsticks agreed upon that measure the inspiration/aspiration of carbon in normal planetary cycles, especially under conditions of changing world temperatures. We do know, with overwhelming certainty, that the carbon cycle that has kept world temperatures within reasonable parameters for millions of years is well out of its historical balance. After demonstrating the existence of that imbalance and demonstrating its implications all the rest is commentary.

Let us consider the mind-boggling numbers when the discussion turns from putting CO_2 into the atmosphere to taking billions of tons of CO_2 directly back out of the atmosphere.

Experimental laboratory technology exists to do just that. There are many suggestions relating to the recapture of released CO_2. One device works by forcing air over a class of absorbing chemicals; then, by heating the material, airborne CO_2 is recaptured. The leap from what can be done on a laboratory scale to what can be done on a planetary scale is daunting. It has been estimated that to counter only the present level of emissions, there would have to be 100 million of these air capture devices working 24/7 to maintain CO_2 at its present level.

To actually reduce the present, inherited level of CO_2 to *preindustrial* levels, there would need to be ten times as many devices, and the costs would be stratospheric. Climatologist James Hansen, Chief Scientist of NASA, and other climate scientists estimate that the cost of reducing ppm of CO_2 by only 50 ppm could be upwards of $20 trillion.[4] Designing, building, maintaining, and replenishing these devices could require up to ten people per device. That is one sixth of the present population of the Earth.

And since energy is required in the capture and sequestration of CO_2, there is no assurance that this method would be warming-neutral as long as the energy required originates, even in part, from the use of carbon-based fuels.

"The cost estimates for capturing CO_2 from ambient air are gross underestimates," says principal research engineer Howard Herzog, of the Massachusetts Institute of Technology. "It's actually still a question whether it will take more energy to capture CO_2 than the CO_2 associated with [fossil fuel] energy in the first place."

Another favorite of the recapture community is the use of algae, which, like trees, take in CO_2 and emit water and oxygen. By one conservative estimate, an algae bloom the size of the Southern Ocean would be required to make a small dent in atmospheric CO_2.

These unthinkable numbers, huge as they are, are only a part of the CO_2 problem. The third world, lacking electrical energy and unable to afford fossil fuels, cooks its meals over simple open fires fueled, for the most part, by wood. Each wood-burning cooking fire emits as much CO_2 in an hour as an average automobile emits in the same time. What is worse, much worse, is that wood fires produce soot. While soot does not remain in the atmosphere as long as CO_2, it is regenerated daily and has a warming effect *seven hundred* times greater than CO_2. One gram of soot has the effect of a 1,500-watt heater operating for one week, and each wood cooking fire emits 2,000 grams of soot each year.[5]

And when we recognize that three billion people live in the third world, and that a significant fraction of that three billion use cooking fires, it might be that the third world exceeds the first as a source of global warming. Furthermore, the warming problems in an industrialized society such as the first world are concentrated in large industrial processes that can be dealt with by cost-effective industrial means. But when you have cooking fires spread across most of Africa and Asia, it is difficult to imagine the mechanism that might rein them in.

4) The Long Life of Climate-Changing Gas

The quantities of these gases in our atmosphere run to billions of tons. However, the actual quantities, bad as they are, are less scary than some predictions of how long these gases continue to affect climate.

University of Chicago oceanographer David Archer is credited with doing more than anyone to show how long CO_2 from fossil fuels will last in the atmosphere. As he puts it in his book *The Long Thaw*, "The lifetime of fossil fuel CO_2 in the atmosphere is a few centuries, plus 25 percent that lasts essentially forever. The next time you fill your tank, reflect upon this."

"If we really do get into a situation where we realize that we've changed the atmosphere too much for our own well-being, there are at least ways to back off of that," argues climate scientist Ken Caldeira of the Carnegie Institution of Washington at Stanford University, an expert on geoengineering. "There's no fundamental limit on how much you could scale those activities up. It's mostly a matter of how many resources you throw at it."

The Caldeira study further found that "regardless of how much fossil fuel we burn, once we stop, within a few decades the planet will settle at a new, higher temperatures." As Caldeira explains, "It [the level of CO_2] just increases for a few decades and then stays there for at least 500 years." He added, "That was not at all the result I was expecting."

The finding was not merely a reflection of some peculiarity in the study, as the same behavior shows up in another, extremely simplified model of the climate, created by Victor Brovkin, of the Potsdam Institute for Climate Impact Research. The Potsdam climatologists found much the same result from much longer term simulations. Their model shows that whether we emit a lot or a little bit of CO_2, temperatures will quickly rise and eventually plateau, dropping by only about 1°C over 12,000 years.

Even more alarming is a release from the National Academy of Science, dated January 28, 2009:

> The severity of damaging human-induced climate change depends not only on the magnitude of the change but also on the potential for irreversibility. This paper shows that the climate change that takes place due to increases in carbon dioxide concentration is largely irreversible for 1,000 years after emissions stop. Following cessation of emissions, removal of atmospheric carbon dioxide decreases radiative forcing, but is largely compensated by slower loss of heat to the ocean, so that atmospheric temperatures do not drop significantly for at least 1,000 years. Among illustrative irreversible impacts that should be expected if atmospheric carbon dioxide concentrations increase from current levels near 385 parts per million by volume (ppmv) to a peak of 450–600 ppmv over the coming century are irreversible dry-season rainfall reductions in several regions comparable to those of the "dust bowl" era and inexorable sea level rise.[6]

It seems convincing that there is no direct relationship between the rate at which we produce greenhouse gases and how long they remain factors in controlling temperature. The present inclination among policymakers, encouraged by the lobbies of the fossil fuel industry, is that by simply slowing the increasing rate of GHGs, the warming process will somehow be halted or even reversed. The studies noted above, and many others, stand in direct contradiction to the optimism of the politicians and the cynicism of industrialists attempting to protect present profits and capital structures. It may well be that even the most heroic efforts, such as eliminating all use of biofuels, would allow us to live only marginally less miserably for some centuries hence. On the other hand, not undertaking heroic solutions seems to guarantee indescribable misery for all life-forms and certain extinction for some.

Because of the ability of carbon to remain in the atmosphere for a very long time, there seem to be no perfect options, even in the longest term, to confronting global warming.

5) Solutions or No Solutions

If we are, by some series of miracles, to haul ourselves out of the mess that we have blindly created by dipping into the inherited lodes of carbon, this is what it would take:

1. RESTRICTION OF POPULATION GROWTH.

The population of six billion will leap to nine billion in less than forty years. We presently have barely enough resources (this is especially true of water) to maintain a level of civility between rich and poor nations. In fact it has been estimated that the world is running at 130 percent of capacity in terms of nonrenewable resources. This is the equivalent of eating one's own liver, a condition that as the supply of liver lessens and the hunger in terms of population increases we will be shortly at a point where, in spite of draconian efforts to avoid world hunger, we will be unable to prevail. With three billion more, adding half again as much to the demand for resources as well as half again as much to the degradation of air and water, the world faces tensions insolvable by present political structures.

Nine billion is only the beginning. At current population growth rates, by the year 2100, there will be upwards of fifteen billion jammed into a hot and crowded little planet. At that level of population we will be living at well over 200 percent of capacity.

It is, however, profoundly unlikely that reason will prevail over sexual and cultural habits deeply ingrained by genes, passions, and religion. The hopeful observation that as standard of living improves, families get smaller is soundly trumped by what has already become a declining standard of living for both the first and the third worlds. For the poor countries it is a very short step between declining standard of living and mass death by starvation. We, not unlike the lemmings, are rushing blindly toward the cliff.

2) END THE USE OF BIOFUELS.

It took twenty years of economic battle with the oil companies to go from twenty miles per gallon to thirty miles per gallon of gasoline in our cars. These interests, arrayed against ending our oil addiction, are now joined by their cohorts in coal, loudly touting the myth of "clean coal." The combined forces of oil and coal, which easily have the economic power to drive and direct our economy, are prepared for a long-term war over a planet that cannot afford the time to defeat them. The natural inclination of invested capital to delay all change is a drag on the economy even in normal times. In other than normal times—such as those in which we now live, with the promise of rising temperatures—delays imposed by the power of established capital are not only threats of disaster, they are existential and mortal realities.

As will be argued later in this book, the author believes that there is a possible solution to the further degradation of quality of life that the continued use of carbon-based fuels will impose. But the central question is not whether the proposed solution will work but whether the balance between the pace of rising temperatures and the pace of the wrenching changes required to mitigate them will tip the scale for survival in time to avoid the worst of what is rushing down upon us.

3) ESTABLISH PROGRESSIVE TAXATION ON ENERGY USE.

Large consumers of energy pay a lower rate than do small users. This is a regressive tax on the energy sector and does nothing to encourage the sensible use of energy. The power produced by fossil fuels is offered at bargain prices even when it is clear that the more energy consumed, the higher the burden imposed on the biosphere.

What is an economy of scale for the large industrial user becomes a diseconomy of scale for the planet. Nothing could be more sensible or easier to apply than a progressive tax on energy, which would finally reveal the hidden costs that high energy consumers have been successfully passing on to the rest of us.

4) ENABLE REFORESTATION.

Earth was once little more than a vast unbroken tract of forest. The capture and release cycle of CO_2 in preindustrial times has been thrown into imbalance by the use of biofuels. In the last few centuries we have been ripping out forests with reckless abandon. As that perfectly sequestered lode of carbon diminishes, a huge price is paid.

Even if we were able to reestablish the total reforestation of the planet, which we obviously cannot do, the amount of carbon in the atmosphere would not be reduced unless we were able to control, in some manner, the normal cycle of forest growth, when carbon is taken up by trees and when carbon is released back into the air.

The natural condition of forests is to be carbon-neutral. The carbon that forests sequester in life is entirely returned by death and rotting. If that were not the case, eons of imbalance would have either fried the Earth or frozen it.

Forests are a temporary sink for a given amount of carbon. The only way to extend the time that carbon is sequestered by forests is to cut trees down and manufacture lumber or some other product with a long and useful life and, at the same time, regrow the trees. The conversion of trees to lumber prevents the return of carbon for the length of time that the lumber is in use. Thus, the only way to approach reforestation, if one wishes to mitigate the release of GHGs, is to abandon the concept of natural forest growth and death and concentrate on controlled regrowth and harvesting of trees. Harvested trees, if put to industrial use as lumber or converted to other stable forms, would effect significant removal of carbon directly from the air.

To work, reforestation would have to be on a global scale and bring back into tree cultivation devastated areas such as the Sahara and other deserts. This would, of course, require global-sized amounts of energy for water desalination to alter local climates — which, on balance, might be CO_2-beneficial, providing that the energy used to desalinate is not derived from burning fossil fuels.

5) RESTRICT USE OF FERTILIZER.

Malthus's logical assumptions that populations press up against the limits of food supply have been perfectly demonstrated by the use of fertilizers. The increase in world population in recent centuries paralleled the leap forward in agricultural output per acre. This increase in food production was further aided by the rapid conversion of forested land to agricultural use.

Since the constriction of our food supply by abandoning fertilizers would be a draconian approach to population control, there is zero likelihood that this would prove acceptable from a humanitarian, religious, or political standpoint. In the not so long run, when population exceeds the ability of even fertilization to keep the supply of food consonant with the future numbers of eaters, we will have to deal with the much more intransigent problem of worldwide hunger.

The worldwide use of fertilizers solves the hunger problem and the overpopulation problem in only the short run. Eventually the increased food supply, the gift of fertilization, will lead to a population that we no longer have the ability to feed. The piper will have to be paid when we no longer have the ability to further increase output per acre.

6) RESTRICT CONSUMPTION OF MEAT.

There is a tale, perhaps apocryphal, of a vegetarian tribe in deepest Africa that forswore meat for 364 days each year and on the 365th day celebrated the Great Meat Guambo, a twenty-four-hour orgy of eating meat.

If the tale be true, that tribe is wiser than the world that exults in a 365-day continuing orgy of meat gluttony. Meat gluttony has repercussions far beyond the hardening of arteries. While too much meat can kill one eater, the vast number of farting, belching, breathing, defecating meat animals are gas factories that are killing the world.

These emissions, exacerbated by animal diets heavy in artificial additives, are a prime source of GHGs in general and of methane in particular. A 2006 United Nations report[7] based on a simple model of the carbon cycle concluded that annual emissions from cattle, buffalo, sheep, goats, camels, pigs, and poultry comprised 18 percent of total GHG worldwide emissions, a higher share than transport. This number seemed bad enough in 2006, but worse news was coming.

In 2009, in a report from the Worldwatch Institute, environmentalists Jeff Anhang and Dr. Robert Goodland argued that previous estimates of GHGs caused by food animals were substantially underestimated. "At least 51 percent of worldwide human-caused greenhouse gas emissions are attributed to livestock," they write. Furthermore, according to the Food and Agriculture Organization, livestock emit about 37 percent of the world's human-induced methane, the most potent greenhouse gas.

WHENEVER the causes of climate change are discussed, fossil fuels top the list. Oil, natural gas, and especially coal are indeed major sources of human-caused emissions of carbon dioxide and other greenhouse gases. But the life cycle and supply chain of domesticated animals raised for food have been vastly underestimated as a source of GHGs, and in fact account for at least half of all human-caused GHGs. If this argument is right, it implies that replacing livestock products with better alternatives would be the best strategy for reversing climate change. In fact, this approach would have far more rapid effects on GHG emissions and their atmospheric concentrations—and thus on the rate the climate is warming—than actions to replace fossil fuels with renewable energy.[8]

Earlier estimations chose to exclude the CO_2 emitted from livestock during normal respiration. On the subject of breathing-related CO_2, while over time an equilibrium of CO_2 may exist between the amount respired by animals and the amount photosynthesized by plants, that equilibrium has never been static. Today, tens of billions more livestock are exhaling CO_2 than in preindustrial days, while Earth's photosynthetic capacity (its capacity to keep carbon out of the atmosphere by absorbing it in plant mass) has declined sharply as forest has been cleared. (Meanwhile, of course, we add more carbon to the air by burning fossil fuels, further overwhelming the carbon absorption system.)

Based on these figures, if we were to change our diets and reduce the consumption of livestock foods by only 25 percent, we would achieve "a 12½ percent reduction in global anthropogenic GHG." By itself this would be as much reduction as was hoped to be negotiated in Copenhagen.

If the actual damage done to the biosphere by the cultivation of meat is 53% as some studies report, it is clear that the fastest and most economical way to reduce GHGs is by way of a meat fast, especially in the developed world. Eating one-fourth less meat than we now consume seems a small and manageable price to pay for a cooler future and, incidentally, healthier arteries.

Conclusion

The likelihood of implementing the six suggestions above stands almost at zero. Each represents a deeply instilled and unbreakable habit developed over eons. More recently these habits have blossomed into a voracious consumption of biofuel-based energy.

It will be argued in subsequent chapters that GHGs cannot be mitigated in time, due to the resistance to change, both governmental and social, inherent in our complicated societies and economies. It is generally agreed that the root cause of warming is the production and consumption of energy. Recognizing that we cannot and will not change our self-destructive habits relating to energy, we will need an entirely new method of the production and distribution of energy to feed these habits.

It will be argued that the use of biofuels will need to be entirely discontinued, and that the only source of energy that is available that might accomplish this is nuclear energy. There are serious problems with nuclear energy that relate to terrorism, waste disposal, costs, and safety, in addition to a widespread antinuclear mind-set. All of these problems, we will argue, pertain not to the use of nuclear energy but to the scale of its use.

Indeed, going beyond the problem of energy generation, it is beginning to look like *all* of the problems of
the twenty-first and subsequent centuries—food supply,
water supply, waste disposal, income disparity, and geo-
political confrontations, to name a few—spring from or
are in one way or another exacerbated by the insupportable scale of projects imposed on our small planet by the
industrial inventiveness of humankind. They are at the
same time too big to fail and too big to maintain.

CHAPTER TWO

What Happened When

The First Nuclear Era

THE FIRST NUCLEAR ERA commenced in 1895, when Wilhelm Röntgen discovered X-rays. This discovery effectively divides all human history into a before-and-after tale. Before the first era, the most powerful force that defined the course of human events was most probably religion. This hegemony was further eclipsed when Albert Einstein described a new power, the power within the atom, which would in a few years become the dominant decider of man's fate on Earth.

The impact came with blinding speed. Barely a decade after Röntgen's discovery, Albert Einstein "disturbed the Universe" by defining the relationship between mass and energy in one simple formula. $E = mc^2$ is shorthand for the physical reality that any unit of mass, a gram or an ounce or a pound of any atom, contains unimaginable energy. In 1938, German scientists Otto Hahn and Fritz Strassmann demonstrated that a splitting process, known as fission, could occur and would be self-sustaining. The discovery that the enormous power of the atom could be released was, to the scientific community, news of world-shattering impact. Niels Bohr, a giant in the history of science, quickly and secretly brought news of developments in Europe to American scientists during the war that was engulfing the world.

For reasons we will develop, these *inherently* safe reactors, are different from nuclear plants that are considered safe by virtue of imposed engineering safeguards. Development of these inherently safe reactors was never pursued even after the near disasters of the inherently unsafe reactors at Three Mile Island and Chernobyl. Engineered safety, as in those two accidents, retained the possibility of meltdown, whereas inherently safe reactors, because of the nature of the nuclear reaction itself, were self-limiting; it was impossible for them to go out of control.

Teller would dominate atomic politics and the science itself for almost half a century, emerging eventually as the nuclear *éminence grise* behind President Reagan and his ill-considered "Star Wars" program. Alvin Weinberg, who remained central in nuclear physics for more than half a century from the inception of the Manhattan Project until his death in 2006, wrote the two definitive histories of the period with seeming total recall. Like Szilard, Teller and Weinberg were chary of engineering safety and in a note to their collaborators suggested that they "take a fresh piece of paper and start all over" on an inherently safe "walk away" design for a nuclear reactor. The names and brilliant accomplishments of the young scientists in the Manhattan Project cannot all be listed here, but most of the individuals involved went on to great careers and great honors in the field of nuclear physics.

When the Trinity test bomb exploded, on July 16, 1945, Robert Oppenheimer, the central driving force behind the Manhattan Project, quoted the *Bhagavad Gita*: "Now I am become Death, the destroyer of worlds." Well, perhaps and perhaps not, as we will argue later in the book.

The antipathy to the use of nuclear power, whether for generating electricity or excising enemies, began even before the war ended, when Hiroshima was destroyed. Among the thousands of scientists who had made the bomb, there were those who, as it became clear toward the end of the war that the Germans were beaten, argued there was no longer a reason for a bomb of this unimaginable power to exist. Germany was no longer a nuclear threat, they argued, and the continuing development and use of atomic weapons had dangerous long-term geopolitical and moral consequences. On one hand the dissenters were bang on target, as the problem of proliferation of weapons-grade material and the seeming ease of building a nuclear bomb were very shortly to dominate international nuclear polity. On the other hand, they were on the wrong side of history, as they could not have anticipated the future pressing need for an alternative to fossil fuels.

Dissent coalesced around a group of scientists at the University of Chicago, where the Manhattan Project was originally based. Szilard, Walter Bartkey, and Harold Urey—the latter directed the project's important research in gaseous diffusion in the search for plutonium—led the group, which was shortly joined by Albert Einstein. This entity was to become known as the Union of Concerned Scientists (UCS). The UCS, along with the fanatically antinuclear Albert Lomis, of the Rocky Mountain Institute, helped set a pandemic public antipathy to all uses of nuclear power, peaceful or military.

Very early on in the development of a self-sustaining nuclear reactor, a most unlikely player appeared at the National Laboratories and demanded that he become a part of the development of nuclear power. Hyman Rickover, a mere captain in the U.S. Navy, was slightly built, arrogant, insensitive to feelings of his peers, and a Jew in an armed service that had the inborn prevailing prejudices of the time. Rickover, described by an admiring peer as a "cold, unrelenting, ruthless workaholic, undermining the bureaucracy while creating his own," was a master at handling senators and congressmen. Most important, he had a nose for any possibility of improving the submarine service of which he was in charge.

In a curious manner Rickover stood astride the developing conflict between large-scale and small-scale nuclear power generation. On one hand recognizing the enormity of the changes that nuclear propulsion was to bring to naval warfare he sought a relatively small reactor to drive his submarines. On the other hand, he was to become an important midwife in the birth of the huge reactors that now provide much of our electricity. Rickover was, unintentionally, the first nuclear mugwump, promoting small reactors for his own use and at the same time pushing for the huge plants that have come to dominate the industry.

On arriving at Los Alamos, Rickover concentrated like a laser on reducing the size of, and standardizing for his own purposes, the light water reactors that had been developed in connection with plutonium breeder programs during the war. He had no need to concern himself with the problems of proliferation that were inherent in LWRs, since his submarine service was a closed, tightly controlled fiefdom. Costs were not a consideration. However, in those days, when the USSR was a looming nuclear power, time was a major consideration. Rickover, always in a hurry anyway, adopted the reactor model that had worked during the war and was on the shelf ready for his almost immediate use. LWRs still had bugs to be worked out. To counter these disadvantages, and to protect his submarine herd, he rigorously trained the men who would operate them. His personal interviews with prospective nuclear submariners became legendary. He was so successful in his training and discipline that the United States has had upwards of one hundred operating nuclear subs for the better part of half a century without a serious mishap. A sizable fraction of American civilian nuclear operators today graduated from Rickover's submarine service.

At the same time Rickover may well be the man who, possibly more than all others, was responsible for America's concentration on the huge light water reactors that comprise today's nuclear power plants. In the early 1950s his reputation as a can-do nuclear engineer attracted the attention of the Atomic Energy Commission in its search for a model for the first commercial nuclear power plant, to be built at Shippingport, Pennsylvania. He proposed, with all of his legendary powers to convince, the form of the reactors that had worked so well for the submarine service for so many years. He became personally involved with every aspect of the project. He was an officer in the U.S. Navy, yet to a great extent he directed the creation of the first civilian nuclear power plant. As a result of his involvement, the AEC adopted the LWR models for which so much engineering had already been worked out. To start from scratch would have been as unthinkable for the AEC as it had been for Rickover when he nosed about the Manhattan Project seeking a mobile nuclear means to power his subs. The die was cast in 1957. Alternative reactor schemes were abandoned, and for almost exactly half a century few in the industry ever even considered developing anything other than gigawatt LWRs.

The first nuclear era came to a close in 1945 with an exultant, all-powerful America secure behind our nuclear exclusivity. For a rare few years the national psyche was untroubled by national security threats from abroad or from disasters inherent in the as yet unbuilt nuclear plants at home.

The war was over due in no small part to America's economic strength. General George C. Marshall's selfless plan rebuilt Europe and then, safe, smug, secure, and a bit too proud, we reveled in our own prosperity and our own unassailable might. These were the few years of innocence before the USSR taught us that there are no secrets.

The Second Nuclear Era

There exists a photo taken in 1951 at the Nuclear Reactor Station in Idaho at the dawning of the Second Nuclear Era. The photo shows four small, dim light bulbs being lit by the building-sized Experimental Breeder Reactor #1. It expresses graphically how far we have come in how few years. This ordinary-looking picture illustrates the tentative beginning of the technological breakthrough of generating electricity from the power of the atom that altered how we live and eventually could alter how we use our fragile planet.

At the beginning of the second nuclear era a split mind-set defined our national nuclear policy. Even as we recognized the energy potential of large nuclear plants, we remained terrified of the consequences of these installations. It was this national schizophrenia that structured nuclear thinking in America in the years between 1945 and 1980. The situation was brought to a head when both pro- and antinuclear camps had to confront their worst fears on March 28, 1979, when the first news of a problem at Three Mile Island was announced.

Around 1951, the AEC chose the LWR as the model that would be used in Shippingport as the first of our major nuclear generation plants. This selection set in stone the pattern for all future reactors in the United States, as it became the sole model that the AEC, and its successor, the Nuclear Regulatory Commission (NRC), would consider for licensing. The research and engineering paid for by the military during the war years led to the investment of many millions more of man engineering hours and more billions of dollars in establishing the ascendency of large-scale LWR nuclear generation.

Shortly after the Shippingport plant opened in 1957, all of the work on alternate nuclear generation methods, such as the use of thorium to replace uranium, came to a halt. Alvin Weinberg, in 1950, with his colleagues at the Oak Ridge National Laboratory, proved on a number of important levels the superiority of self-limiting thorium reactors in hundreds of tests. But the U.S. military, in response to the growing nuclear threat from the Soviet Union, was hungry for the weapons-grade plutonium that LWR breeder reactors generated as a by-product of their thermal output. In the nuclear panic of the time that engulfed the United States when we learned that the Russians had the bomb, the military could not be denied. This intervention into the measured progress of civilian scientific research by the military, with its immediate concerns for national safety, guaranteed that the course of the nuclear power industry was set for the next four decades.

The use of inherently safe reactors for energy generation, suggested by Teller and pushed hard by Weinberg, became one of the great what-if technologies of the twentieth century. It was not until the turn of the millennium, in response to an equally deadly threat of carbon in the atmosphere, that the books were opened once more on alternatives to light water reactors. However, when the world began to recognize that nuclear power was the sole logical substitute for biofuels burned in our generating plants, the industry and the regulators chose to accept the limited real risks of LWRs over the imagined possible risks of what was, to them, a novel technology.

Developments in LWRs burgeoned in the decade of the 1950s. Especially interesting was the development of relatively small thermal output nuclear reactors such as were developed by then captain and now Admiral Hyman Rickover for use at sea. The first American nuclear-powered merchant ship, the NS *Savannah*, was launched in 1962. The Soviet Union had launched its nuclear ice breaker in 1962, and, of course, Rickover had been churning out nuclear subs as far back as the 1950s.

The *Savannah*, judged by some the most handsome merchant vessel ever built, remained in service until 1971 when it was deemed too expensive to operate compared to ships powered by oil. Eight years later, as the price of oil increased, the *Savannah* would have been more economical, but she was never put back into service, probably because maritime officials around the world, led by the Japanese, still smarting from Hiroshima, closed their ports to any nuclear-driven vessel. In fact, at 2010 prices, reactor fuel costs were 85 percent less than bunker oil. The primary users today of nuclear propulsion at sea are the U.S. Navy and the U.S. Submarine Service. To propel one of our giant aircraft carriers at 25 knots for twenty-four hours requires only three pounds of cheap nuclear fuel. Great Britain and Russia also maintain fleets of nuclear vessels.

The last mobile nuclear power plant built by the U.S. Army for civilian use was installed on a converted liberty ship, the USS *Sturgis*. The plant was built to supply power to the Panama Canal, which it did from the year it went critical in 1967 until it was shut down in 1976. Of particular interest was the small size of its output as it generated only 10 MWe (megawatts electrical). It was designed to be sailed to wherever power was needed. One might conjecture that the plant was built aboard a ship to obviate the attendant risk of proliferation of its falling into the hands of the Panamanian government.

Curiously, no record of the nuclear use of this vessel can be found except for a few lines describing its capabilities published by the Energy Information Administration, whose job it is to generate energy statistics for the government. It looks as if information about the USS *Sturgis* as a nuclear power plant is still considered secret. However, no one seems to have notified the EIA.

Also of special interest, in the context of this book, is the ML-1, a "reactor in a box" that first went critical in 1965. It was a farsighted proposal by the U.S. Army, except that it did not work out very well due to a very difficult engineering profile. However, it did make electricity for about one hundred hours and demonstrated that small, transportable nuclear reactors were well within the technical capabilities even at that early date. The complete ML-1 system required six "boxes," not one. In addition to the reactor box, a container was needed for the control room, another for the heat conversion system, and a total of three boxes for the cabling, auxiliary gas storage, and handling equipment, and tools and supplies. The containers could then be loaded aboard a train, truck, or plane. Protection against radiation was handled in a novel though logical manner: operators were required to stay 500 feet from a functioning reactor.

It was, according to Adams Atomic Engines, Inc., "the first nitrogen cooled, water moderated reactor with a nitrogen turbine energy conversion system." Like other projects of the moment, the ML-1 fell victim to the money needs of the Vietnam War. Too bad. With some additional R&D and a redesigned energy conversion system, the ML-1 could have speeded up the acceptance of the idea that nuclear reactors need not be unsafe gigawatt monsters, with all their problems of scale.

On March 16, 1979, Hollywood released *The China Syndrome*. This immediately popular film tapped into the cold war terror of anything nuclear. For thirty years, since the end of World War II, well-intentioned antiwar groups had proliferated and propagandized the horrors of a nuclear war. During the same period antinuclear groups, more sharply focused on peacetime nuclear reactors, were hawking scenarios of nuclear power plants going rogue. The opposition to both wartime and peacetime deployment of nuclear power developed a massive following in the United States. It is no wonder, then, that twelve days after the release of the film, when there was a partial meltdown of the number 2 reactor at the Three Mile Island (TMI) nuclear plant ten miles outside of Harrisburg on the Susquehanna River, a population already nervous went ballistic.

The panic that followed the botched release of misinformation about the true conditions at TMI was enormously more damaging than the insignificant release of radioactive material relating to public health. However, it did propel the Hollywood version into blockbuster status. The film, magnifying the reports emerging from TMI, helped the titer of national panic to leap beyond the facts. The eventual effect on the development of nuclear power in the United States was a blanket resistance to the construction of nuclear plants for the next twenty years. This was not a significant problem at the time, since we had plenty of coal to burn. But the accident at TMI and a nonscientific fictional Hollywood film set us back two decades in the coming war against global warming. Global warming, enhanced by the continuing burning of fossil fuels, was the real threat rushing down upon us, the consequences of which, we are now learning, are much worse than serious, they are deadly.

As if that were not enough to fuel an antinuclear pandemic, reactor number 4 at Chernobyl went out of control on April 26, 1986, and Europeans, downwind of the reactor joined the hue and cry against nuclear power. The eventual costs to Russia in terms of dislocation of population and cleanup were enormous. However, the number of deaths directly attributable to the Chernobyl event were fewer than sixty. By comparison, deaths in the same period directly attributable to energy-related industries *other* than nuclear totaled more than fifteen thousand.

According to the United Nations, Chernobyl will eventually be responsible for some four thousand indirect deaths over the lifetime of the affected population. Compared to the worldwide indirect loss of life in the same period from nonnuclear energy industries such as coal mining, oil drilling, and hydroelectric dams the number is, relatively speaking, small. More specifically deaths exclusively from the pollution of our atmosphere as a result of burning coal and oil number in the hundreds of thousands each year.

One statistic unrelated to energy generation brings home with force the impressive safety record of the nuclear industry. From 1979 until this writing, over half a million deaths have resulted from lightning strikes. One is 8,000 times more likely to be killed by lightning than by a nuclear accident.

Both at Chernobyl and Three Mile Island, the systems worked as they should have in emergencies, but the actions taken by those who operated the systems— because of lack of training or because the operators were misled by the inadequate and unsophisticated controls of the time—created the problem. The Chernobyl incident started with a routine shutdown of the plant to test functioning at low power. Previous Russian experience had reported that this was an unsafe procedure but this was evidently unknown to the Chernobyl operators. The disaster would have been prevented had there been better internal communication.

The Three Mile Island event was precipitated by a minor malfunction in a secondary cooling circuit that caused the reactor to shut down within milliseconds.

The problem then occurred when a relief valve failed to close. The result was reminiscent of Chernobyl: "The operators were unable to diagnose or respond properly to the unplanned automatic shutdown of the reactor. Deficient control room instrumentation and inadequate emergency response training proved to be root causes of the accident."[10]

False perceptions concerning nuclear accidents remain in the public consciousness. As noted, the death toll at Chernobyl was small, and the fire did not destroy the entire installation. In fact, three of Chernobyl's four reactors were later returned to service. In TMI-2 there were no fatalities. Its sister reactor, TMI-1, later completed the longest operating run of any light water reactor in the history of nuclear power worldwide—616 days and 23 hours[11] of uninterrupted operation. As often happens, however, failures are more newsworthy and longer remembered than successes.

These events marked the end of the fifty-year-long second nuclear era, during which almost a hundred reactors in the United States generated power with remarkable safety. The net effect of the delay in the construction of additional nuclear plants between 1980 and 2010 was to make a substantial contribution to global warming during a period when an expansion of nuclear power might have been most effective in diminishing the CO_2 being poured into the air. The panics precipitated by Three Mile Island and Chernobyl also shut down the first tentative steps to create a nuclear merchant marine in spite of flawless operations at sea by the U.S. Navy.

The odds were all irrationally stacked against anything nuclear. It would take the developing clear and present danger of increasing global temperatures and rising sea levels and all of the ills descendant from the soaring tonnages of greenhouse gases to dent, but not yet destroy, this opposition. Public opposition to nuclear power along with the disinclination of the regulatory agencies to gear up to new challenges are the hurdles yet to be overcome in the twenty-first century.

CHAPTER THREE

The Trouble with...

ALL ENERGY SOURCES available on Earth save nuclear and geothermal[12] are powered by the sun.[13] These are the various ways that our planet has chosen to use and store the sun's energy. All of these energy sources add their mite, and sometimes their might, to the imbalance between the energy the Earth receives and the energy it sends back out into space as all of these energy sources are including geothermal emitters of CO_2.

Each form of usable energy that derives from the sun has a bundle of characteristics that define its advantages and its drawbacks in a complicated industrial society. The profile of any resource includes considerations of geographic distribution, specific energy content, difficulty or ease of conversion to usable electrical energy, depletion rate, and agility of response to commercial marketplace variations.

Add to these the formation and deformation of CO_2, both natural and man-made, that is common to all energy use that derives from fossil fuels. It is this human-driven process that is shaping humanity's future on Earth. It is the deleterious, existential effect that *any* fossil fuel energy source has on the environment. Some systems of energy generation, such as coal and oil, emit enormous quantities of CO_2. Others add lesser fractions. But since rising CO_2 levels in the atmosphere are predicted to change our climate forever, any ounce of that gas that does not enter our atmosphere is an assist to the comfort and continuity of the human population, as we now know it, on Earth. And any ounce that is allowed to enter our atmosphere will add to the deep discomfort of our children and grandchildren.

The most important factor that defines the CO_2 matrix is the rate of energy release. Fossil fuels were created over millennia, yet the increasing release of these millennial troves of fossil fuels and resulting release of greenhouse gases are now being measured in mere decades. The Earth has many mechanisms that maintain a balance between sequestered CO_2 and the amount in the atmosphere, but these mechanisms that control the volumes of CO_2 held in the seas and sequestered in the earth itself can deal only in incremental, slow, planetary velocities. There are no natural mechanisms that can cope with the CO_2 that has been let loose in just the past two centuries primarily through the burning of fossil fuels.

Solar and wind energy systems, while having a less deleterious effect on the environment than fossil fuels, are examples of electrical generation with a minimal rate of release problem. Energy from these and similar resources is captured and used as it is created and in the amounts that can be instantly injected into a grid system. However, as will be seen, solar and wind generation also create greenhouse gases on a minimal but significant scale.[14] These sources call up no ancient millennial lode of carbon to be dumped into our environment. However, such lesser sources of energy still contribute an unacceptable burden of CO_2 into our energy matrix.

Oil, coal, and natural gas, compared to wind and solar, are a very different matter. Their use is analogous to a dam, in this case backing up eons of tons of carbon, that suddenly gives way. The planet, demonstrably, cannot in so short a time absorb these by-products, which were billions of years in the making. It is being overwhelmed by the high and wonderfully useful specific energy content of fossil fuels. As we drew on this embarrassment of riches during the past two prodigal centuries, we could not foresee our future need for energy, nor were we able to control our voracious appetite for it.

In an economy that takes into account little other than short-term profits and gratifications, the true future costs of biofuels impose no rein on their use. If all the costs and outcomes, present and future, of the use of oil and coal and gas were calculated, these fuels would only be used and priced proportionately to the ability of the planet to absorb their consequences. The price of fossil fuel that would have emerged might well have been two orders of magnitude more than we now pay for energy. Unfortunately, the market-driven engine of our economy needs to demonstrate immediate profit, which results in severe discounting of future costs. As a result, conservation and proportionate use, long-term considerations that are enemies of immediate profit, have little weight in commercial decisions.

Each of the score or so of sources of energy in use today has a deleterious effect, some more than others, on the warming problem. Because of the enormous backlog of CO_2 that has been and is being pumped into the air, every energy source that produces CO_2 needs to be identified and eventually eliminated if we are to somehow mitigate, let alone reverse, the coming alterations in our planetary climate.

Thus, what follows here[15] is about winkling out the hidden costs and the hidden sources of every stray ounce of CO_2 and methane that emerges from the production of energy, no matter how seemingly inconsequential in the big picture any might be.

Recapture from Air

"If we really do get into a situation where we realize that we've changed the atmosphere too much for our own well-being, there are at least ways to back off of that," argues Ken Caldeira, an expert on geoengineering. "There's no fundamental limit on how much you could scale those activities up. It's mostly a matter of how many resources you throw at it."

There exists at laboratory scale a device that can suck CO_2 directly out of the atmosphere. The first problem of this method is the scale. It could take millions of large devices operating on a continuous basis to make a dent in the amount of CO_2 in the atmosphere, and it does not address the methane problem at all. But although this process is intended to be carbon-negative, much depends on the amount of CO_2 produced by the manufacture and installation of the devices and the millions of tons of absorbent chemicals used in the capture.

On the matter of scale, Klaus Lackner, of Columbia University, reports:

> A device with an opening of one square meter can extract about 10 tons of carbon dioxide from the atmosphere each year. If a single device were to measure 10 meters by 10 meters it could extract 1,000 tons each year. On this scale, one million devices would be required to remove one billion tons of carbon dioxide from the atmosphere. According to the U.K. Treasury's

Stern Review on climate change, the world will need to
reduce carbon emissions by 11 billion tons by 2025
in order to maintain a concentration of carbon dioxide
at twice preindustrial levels.

The process requires that air be pumped across
absorbing chemicals that need to be heated to emit the
captured CO_2. Depending on the source of the energy to
pump and to make the heat required to heat the chemicals, the CO_2 produced by these activities must be
deducted from the total CO_2 captured.

Another overriding problem with this method of
recapture is that the hours required to design, build,
maintain, resupply, and refresh the millions of devices
might be upwards of hundreds of millions. Further is
another problem of scale, not unique to direct carbon
capture: Where would we sequester a billion tons of CO_2
a year for at least the next decade? Where would we get
the energy to do so? And how much CO_2 would be emitted in the process of sequestration?

Essentially, air recapture, because of its scale
alone, would "require the creation of an industry that
moves material on a scale as large as (if not larger than)
that of current fossil fuel extraction, with the risk of
substantial local environmental degradation and significant energy requirements."[16]

This is geoengineering with all of the unintended
and frightening consequences that emerge from mess-
ing about on a planetary scale.

The final word on the subject comes from David
Archer, who sees the prospect of recapture as an exer-
cise in futility compounded by stupidity.

"The only geoengineering scheme that deals with
the persistence of CO_2, is to extract CO_2 from the atmos-

phere, to really clean it up....The problem is that once CO_2 is diluted by releasing it into the atmosphere it takes energy and work to unmix it. Releasing our fossil fuels CO_2 to the atmosphere now is a phenomenally stupid strategy, if the eventual plan is to clean it back up."[17]

Every pound or ton or ounce of CO_2 released now will require enormously more resources in the future to remove it. From the point of view of recapture, the only sensible and sustainable scheme is to resist compounding the recapture problems in the future.

Algae

As we have seen, algae, like trees, can take in CO_2 from the atmosphere and release water and oxygen. It has also been demonstrated that algae can be harvested to create a livestock feed, thus expanding the human food supply. Additionally, algae can be processed into a bio-fuel that has the ability to replace fossil fuels.

Algae is thus proposed as a magical means of sequestering carbon, providing feedstock for food animals, and as a substitute for fuel oil. Let us look at these claims.

Algae are living beings. They are capable of breathing in and retaining carbon. This carbon becomes the corpus of the animal for as long as it remains alive. When algae die, or indeed when we die or anything dies, the rotting that takes place absorbs oxygen from the air and reinjects into the air the temporarily sequestered carbon.

This would mean that when an upper limit of sequestration is reached, that limit being the amount of surface water dedicated to growing algae, the only way to continue the process would be to harvest all of the algae of that cycle and let a new growth appear. That would appear logical and beneficial, except that this is only the middle of the process; something would have to be done with the carbon sequestered in the remains of the first and subsequent cycles. The harvested algae would have to be prevented from rotting and find a permanent use.

One proposed use would be to turn algae into livestock feed. Food animals, like most living things, are nothing more than engines for the manufacture of CO_2. We will thus be going through a long, expensive, and unnecessary cycle of removal and replacement. Ultimately, the same amount of carbon that had been sequestered in the atmosphere and removed by algae will appear again in the atmosphere as greenhouse gases emitted from the front and rear openings of food animals.

Why bother?

An even worse proposal is to reconfigure algae as a biofuel to replace fossil fuels. In this feckless event the carbon, now a burnable oil, need not go through the body of a steer; it goes directly up a power plant chimney and back into our burdened air.

The illogic of such a fruitless cycle seems to be missed by the fossil fuel–burning industry. One company, Alabama Public Service, upon demonstrating at an experimental level that algae could replace oil, declared, "The results provide evidence of the financial viability of using the emissions of a power plant to grow algae for the exclusive purpose of creating biofuels." This service is seriously proposing that we take the carbon from the smokestacks, convert it to a fossil fuel, and then burn it so that the carbon goes up a different smokestack. The net effect of such a process is to increase the amount of CO_2 in the air since the process of making a fossil fuel from the CO_2 the results from burning coal itself requires energy from burning coal... *ad infinitum.*

The scale of algae production that would be required to remove a significant amount of the CO_2 that we are dumping into the air would indeed be heroic. A United Nations committee has estimated that, in order to take up only one-eighth of the fossil fuel emissions that we annually produce, it would be necessary to "treat" the entire surface of the Southern Ocean. And we would have to repeat this treatment every year.

Picture the armadas of ships required to treat and to reap the algae from the 20 million square miles of this sea. And then what would they do with those billions of tons of material loaded with CO_2?

Mark Edwards, a passionate advocate of the algae solutions, curiously reminds us: "Over the last century many people and companies have lost fortunes trying to create commercial scale algae production. The laboratory studies are so promising, yet even on modest scale field studies have typically become unmanageable, unstable and unproductive."

Biochar

If we lived in a rational world where nations, corporations, and institutions agreed on specific concerted actions and if—a very big if—we had the leisure of half a century to respond to global warming, the universal use of biochar might be seen to have a serious benefit. The question mark is still there, however, because the beneficial effects of which biochar is capable have as yet to be proven outside the laboratory.

Biochar is created when any mass of biomass, that is, anything that uses carbon in its makeup, is heated in an atmosphere free of oxygen. The process is called pyrolysis, and the effect of the burning is to sequester in "char" the carbon previously contained in any mass of agricultural wastes or most other wastes. Additionally, biochar increases the fertility of most soils and is also beneficial to agricultural soil in that it retains water.

This process raises myriad questions. First, all of the research on the long-term predictions of massive reliance on biochar has been on a laboratory scale. To be effective, the process would have to go planetary, and that raises the specter of unforeseen climatic consequences that might be directly antithetical to our intent. One suggestion is to grow billions of acres of tree plantations as a source of biochar. This again is geoengineering on a grand and dangerous scale.

As the disastrous impacts of industrial plantations for pulp and paper and for agrofuels have already shown, land conversion on this scale poses a major threat to biodiversity and ecosystems, displaces communities, interferes with food production, and degrades soil and freshwater resources. The proposed use of "agricultural and forestry residues" is based on unrealistic assessments of the availability of such materials, the removal of which deprives soils of nutrients and organic matter, encourages erosion, and reduces critical habitat for biodiversity....The risk of severely worsening rather than mitigating climate change exists if emissions from land use change or if unanticipated soil carbon or "biochar" carbon losses occur.[18]

From the point of view of this book, which argues that *any* creation or release of CO_2 is eventually more damaging more expensive and longer lasting than any sequestration of already released CO_2, biochar fails on at least three levels.

In order to make biochar, the biomass that is used must be heated by burning. If any source of the heat is other than nuclear, there will be a net addition to atmospheric CO_2. The proponents of biochar themselves admit that less than 50 percent and sometimes as little as 20 percent, depending on the makeup of the biomass heated, of the original carbon is sequestered in the process. One must ask, and the question has not yet been answered, where the remaining carbon goes if it is not sequestered. If this missing carbon goes into the biofuels, thus becoming a by-product of pyrolysis, and these fuels are subsequently burned in energy production, then it seems that we have traveled a long, tortuous, and expensive route. The use of biochar inevitably ends up adding to the CO_2 problem, and it is also likely to add to the bottom line of private industrial and agricultural corporations, the very institutions that are creating the CO_2 problem in the first place.

What is wrong with this picture?

Generally what is factual and proven to be known about a technology becomes the basis for assessing the risks of applying it. In the case of biochar, it is what is unknown about this untested technology that raises the red flag of unforeseen and unacceptable planetary risks.

Coal

The burning of coal is one of the chief sources of CO_2 in the atmosphere. This would not be so pressing a problem except for characteristics that, taken together, uniquely define coal and its impact on the environment.

The first characteristic is that of vast, overwhelming quantity. There has been a tremendous amount of coal accumulated over the ages by the deaths of countless life-forms. Because of the huge supply, the danger derives not only from what coal emits when it is burned but rather from the scale of the burning. The rate of release of inherited resources of millennia creates an imbalance within normal geological cycles and natural carbon sinks. To make matters worse, predictions[19] indicate that the world's coal consumption will increase by half in the near future, a reflection of population growth. Even at its present and accreting rate of usage, coal is still available as a cheap and abundant source of energy. There is enough coal in the ground now, one trillion tons, to continue to pump billions of tons of CO_2[20] into the atmosphere each year for the next two centuries at the current rate of consumption.[21]

The second characteristic is coal's ubiquity. No other useful energy resource is as widely available in the world as is coal. Commercially exploitable deposits, at current price levels, occur in Europe, Asia, North America, and Australia, essentially throughout the industrialized world. The fact that coal beds are so widely dispersed reduces transport costs and makes coal more economically desirable as a fuel than, for example, oil or natural gas, both of which emit CO_2 in their use as well as in the transportation required from geographically limited distribution. When coal is immediately at the hands of people who are cold and need power, there are few arguments that will mitigate its use. International concerns, as an American president once wisely opined, are local matters.

The third characteristic is its ease of use. Coal is dug out of the ground, chopped up or pulverized by simple and cheap industrial processes, and consumed by each and every generating plant at the rate of a hundred tons every twenty minutes. Oil must be refined and shipped, natural gas must be compressed and shipped, but coal, widely underfoot as it is, needs little more than a shovel and a diesel-driven wheelbarrow.

A fourth characteristic related to the third is accessibility. While coal lies in convenient and easily mined seams underground, a more recent method simply involves blowing the tops off of mountains enabling open-pit mining by gargantuan earth movers, the cheapest and quickest way to get at the coal. Big Brutus, not the largest of these machines, moved 135 tons of coal at each swipe.

And the fifth characteristic is its efficiency. Coal is a superb source of heat energy; it burns easily and hot. Due to its availability and cost, 75 percent of all coal that is mined is dedicated to coal-fired generating plants, the prime source of greenhouse gases in the world. (As an example, among the five top CO_2-producing coal-fired power plants in the United States, each emits over 20 million tons annually.) Coal delivers cheap electric energy to a developed world that is addicted to its cheapness and to an underdeveloped world that desperately wants to become addicted.

There are thousands of electric utility firms, making this one of the largest industries in America.

These mostly private companies crank out roughly 75 percent of the nation's power from coal-burning plants that are scarcely more energy-efficient than they were in the 1960s. To increase plant efficiency and reduce CO_2 output, the United States passed the Clean Energy Act, but at the insistence of lobbyists for this politically powerful industry existing plants were exempted from the regulatory process. As a result, utility companies had and still have little impetus to reduce CO_2 emissions or to build more efficient plants. Indeed, the net effect of the Clean Air Act was to extend the life of CO_2-spewing power plants for decades. It was cheaper for utilities to preserve polluting old plants than to pay the price in regulation when new plants were built.

A growing number of coal-burning power plants around the world, in response to public concern over global warming, are incrementally reducing their emissions while creating another problem: water pollution.[22] As coal-fired electric generating plants continue to use up our planet's ability to adjust to increasing CO_2, these plants are at the same time using huge quantities of water to process their coal,[23] thus depriving millions of the water without which a modern industrial society cannot survive. We are simply trading cheap electricity for water that is becoming more scarce and more expensive each day. Some experts predict that, in the immediate future, potable water will be a more exacerbating problem than even global warming. For the most part, warming is taking place on an extended time scale, although some effects will be immediate, while water shortages are already with us today, here and now.

As our hunger for electricity[24] from coal plays out, it becomes more and more difficult to conceive of the

Conservation

The statement the less energy you use means that you use less energy seems to be a tautology of a high order. The argument that saving energy saves energy is the mantra of those who offer energy conservation as a no-cost method of reducing energy use.

The best way to conserve energy is not to use it. Energy not used, as when a lightbulb or a computer is turned off when not needed, is a true saving in overall consumption. It is a stand-alone, irrefutable statistic as long as one assumes all other consumption remains the same. Conservation achieves wonders in reducing our dependency on electricity no matter how it is generated or at what cost. Under static conditions, the cost of energy that is not used is essentially zero, so it would seem that we are not only reducing greenhouse gases but also saving money. This is a fact but it is not a *true* fact.

The true fact is that as soon as conservation is looked at in the bright light of a dynamic and ever-changing economy, conservation as a result of increased efficiency no longer leads automatically to the saving of the total amount of energy used. It is not enough that the cost of the energy not used be zero to the user. In order to effect a saving, the cost of energy not used must be less than zero in the matrix of an every shifting free economy.

Stanley Jevons was an economist and philosopher of science who lived in the 1800s in England. Jevons was among the first to address the cost of energy not used. He observed that when James Watt's new engine improved the efficiency over existing steam engines by a factor of five, the use of coal, which logic would decree would fall, instead rose dramatically. In 1865, in an article dealing with England's coal reserves, he suggested what has come to be known as the Jevons paradox. The paradox states that greater efficiency in the use of an energy source almost inevitably leads to greater consumption of that source. The paradox denies the relationship between increased efficiency and decreased consumption. What seems to be a logical relationship, paradoxically, is not.

When Watt's new steam engine was shown to use a small fraction of the amount of coal to do the same job that, for example, the earlier Newcomen engine used, an enormous new market for energy use opened up. While each steam engine now used less coal, the total usage of coal in England soared. The decrease in energy cost enabled manufactured products to become cheaper, and, as a result, demand unconstrained by high prices soared, as did the consumption of coal.

In fact, in the past hundred years, during which the efficiency of burning coal went up incrementally, it was matched by an exponential increase in the total consumption of coal, a classic proof of the Jevons paradox. For each ton of coal we used in the mid-nineteenth century, by the mid-twentieth century we were using approximately one thousand tons.

There is a host of examples that support the Jevons paradox, but there is one proposal that casts a cold light on the "cold lights" otherwise known as compact fluorescent lightbulbs (CFL). In 1987, a forward-looking electrical utility in Traer, Iowa, offered to exchange incandescent bulbs for the 80 percent more efficient CFLs. During the course of the project, 18,000 CFLs were distributed. Contrary to all expectations, total energy use went up in spite of the fact that the incandescents that were replaced had consumed four to five times more electricity than the new CFLs. Electricity consumption in those households that were now using CFLs went up 8 percent. The answer to this conundrum was simple. Householders, aware that they were burning fewer watts, kept their lights on longer.

In 2007, in a policy statement on energy efficiency, the International Chamber of Commerce stated that efficient energy use "reduces emissions and other environmental impacts" and "extends the availability of large but nonrenewable *resources.*" Jevons and the town of Traer and others bring the first statement into doubt. The second statement defines the worst possible outcome in a world already awash in CO_2, as extending the availability of nonrenewable resources would serve best to keep prices down and consumption up, and would keep the rush of carbon into our atmosphere even further beyond the present ability of the Earth's natural ability to compensate.

It is a direct result of the increase in use of biofuels in the past century that we are now faced with the warming of the planet. While the per-unit cost of energy decreased, the resulting enormous costs that we will face in the next hundred years to reverse the impact of those cheap energy years was not factored in. The cost of the standard of living that we have achieved by the profligate use of energy may well exceed the cost of maintaining it by many orders of magnitude. Indeed, it can be argued that, considering the wrenching changes that would be required even to mitigate global warming, it is debatable that our present high standard of living can ever again be enjoyed in any future industrial society.

Also, the argument that decreasing energy costs "extends the availability" of energy reserves is contrary to fact. Proven reserves of biofuels as a whole have been declining due to excessive use and exploitation, which would be exacerbated by increasing availability. While we are not yet at a moment when we can reliably predict when biofuels will become too scarce and too expensive for use as an energy source, we are approaching those price levels. Even now the costs of oil, coal, and natural gas are at high historical levels. With the advent of soaring demand from previously underserved consumers in Asia and Africa and from the predicted requirements of the battle against global warming, there is no reason to assume that the upward surge in the demand for energy will diminish. At some point we will have to pay for all of the perceived benefits of our past profligacy. Restraining price increases now by increasing supply seems to be an Alice in Wonderland solution. We should, instead, be decreasing supply and increasing price.

Benefits to be had from the increasingly efficient use of energy cannot come from within a system, since each consuming economic unit will rationally attempt to maximize the short-term benefits that emerge from more economically available energy. The total effect will inevitably be increased use. The only way that the benefits of efficiency would lead to a decrease in energy use would be by fiat from above such as was the case in Romania under Ceausescu.

In the years leading up to his ouster, the Romanian dictator decreed that no household in the country could be lit by more than one forty-watt bulb. The point was to conserve energy; punishment for be useful, cold air under pressure must be mixed with natural gas in order to become yet another fossil fuel to drive a conventional turbine. A process intended to smooth the grid must use natural gas on the exit, which means that there will be a release of greenhouse gases into the atmosphere in both the compression and the decompression modes of CAES.[25]

A further problem, still little understood, is that air stored in natural underground caverns, as is being proposed, is likely to absorb material from the walls of the storage site. Since most rock contains CO_2, methane, and other egregious gases, it is likely that the air coming back out of compression and being released will add its own greenhouse gases to the atmosphere.

Another well-known process for storage of energy is to lift water or dam it at height. As will be shown later in this book, even hydroelectric energy stored naturally by rainwater in a man-made lake is a source of greenhouse gas. Excess energy from any source that depends on water pumped to a higher level for later use to create energy will, due to the emission of CO_2 from the lifting energy, not only produce a greenhouse gas but result in a net loss of total energy in the system, due to built-in inefficiencies of the pumping process.

On a smaller scale, which perhaps sheds light on the process of water storage, the author has an underground cistern that is replenished by rainwater from the roof. To be useful, the stored water must then be pumped up to height to create pressure for watering and car washing. While saving water, even this small system adds greenhouse gas in the use of the electrical energy required to raise the water to a useful height. On the other hand, if rainwater is caught and stored at height and the fall is used to provide pressure for household water, no greenhouse gases are produced. A glance backward reminds us that this is the way the Romans did it.

There are a number of other proposals for storing excess electricity, including the use of the ubiquitous lead acid storage battery, the spinning of flywheels,[26] and the manufacture of explosives. These and other proposals, all of which produce greenhouse gas, hardly rise to the scale necessary to make them economically feasible in a system that annually produces tetrawatts of energy and megatons of CO_2.

Forests

Eight to ten thousand years ago our ancestors, tiring of the endless exertions and dangers of a vagabond hunting society, observed that growing food was much less strenuous than chasing it across the veldt. So they invented agriculture, which, as William Ruddiman observes, is "the largest alteration of Earth's surface from its natural state that humans have yet achieved."[27]

It was this profound and, in planetary terms, sudden change in man's method of sustaining himself that initiated the first rise in homeogenic CO_2 in our atmosphere. While the measure of the temperature of the Earth is immensely complicated, the measure of the growth of CO_2 in the atmosphere is as simple as discovering long-buried ancient ice-bound bubbles of air and analyzing them. While we are presently wrestling with the impact of fossil fuels, we must look back at what was an inherited preexisting sink of CO_2.

Land has been denuded of forests for eight thousand years in a fairly linear fashion by farmers. That provide refuges of wilderness that all of us humans seem to require at one time or another.

While each pound of CO_2 removed from the air by harvesting forests is a net plus for human life on Earth, the end of forests may well lay a heavy emotional cost on us, and an even heavier cost on all of the other life-forms with which we share the planet.

Geothermal Energy

The golden goose of "free" energy contained in the limitless internal heat of our planet is being severely wounded by the unintended consequences of its extraction. Damaging consequences appeared as soon as matters emerged from the laboratory into the real-world dimensions.

While events can be either controlled or eliminated in the laboratory, it proves not so easy when novel technologies, not quite fully developed, confront the Earth's resistance to significant change.

This is doubly depressing in light of recent rosy predictions of the amount of geoenergy available to an energy-greedy world. One such prediction, suspect but modest, from the Geothermal Energy Association, claims that "Ten gigawatts (GW) of U.S. geothermal power appears to be feasible in the near to mid-term." That amount is equal to the power from ten large nuclear plants, or about 20 percent of the power requirements of California. Considering some recently emerging problems, the rosiness of this prediction lies in its claim to "near term" availability, a serious plus in our coming battles with a rising level of CO_2. At the other end of the scale is the prediction that geothermal power in the United States could produce up to *60,000 times* the country's current annual energy usage.[28]

Although it is true that just under our feet, some miles down, there is enough heat to satisfy the wildest dreams of energy avarice, getting that energy onto the grid without doing collateral damage to a warming atmosphere is becoming progressively more difficult. One example is the remarkable inefficiency of a geothermal plant in Utah, in operation for six months. The best it was able to do was to sell 5 MW of electricity to the city of Anaheim while buying 4 MW of electricity from Rocky Mountain Power's fossil fuel–fired plants. This plant not only failed to repay its investors and reduce our carbon burden, it increased the carbon burden of the industry. To be fair, this is not typical of all geothermal start-ups, but it demonstrated how not to build energy sources to reduce CO_2 output.

Other geothermal plants face problems that energize public opinion against them. Since no geothermal plant has been built or will be built without taxpayers' money, public opinion may be the short-term death knell for this energy source. One event that occurred over six centuries ago is still defining opposition to seeking heat from the heart of the Earth.

On October 18, 1356, an earthquake leveled the city of Basel, Switzerland. According to historical records, "no church, tower, or house of stone in this town or in the suburb endured, most of them were destroyed...hundreds to thousands perished."

It is no wonder, then, that fear approached panic in modern Basel when, in January 2007, it was hit by an earthquake that registered 3.6 on the Richter scale. The quake occurred precisely at the site of a drilling project that was accessing a thermal anomaly under the city. Since it is widely believed that geothermal drilling has produced damaging earthquakes elsewhere, the civic memory of the citizens of Basel fixed on the geothermal plant as the culprit. In at least two instances, one in Landau, Germany, and this one in Switzerland, the fall-out from human-generated earthquakes has resulted in considerable monetary loss. Both plants were forced to shut down.

But what may be insupportable are liability costs, civil and criminal. Markus Haering, the CEO of the firm that was doing the drilling in Basel, was accused by Swiss prosecutors of deliberately causing landslides and damage to property. He denied all charges and was acquitted in 2009. Nonetheless, it is one thing to be able to operate safely behind corporate shields, with the punishment little more than a fine, but when prosecutors start to discuss years in prison for individual employees of businesses involved in geothermal retrieval, then things are seen in a new and more searching light.

AltaRock Energy, another firm pursuing an advanced geothermal energy technology, had to suspend its first attempt to drill a deep well in northern California. After expenses totaling nearly $17 million, some of which came from the U.S. Department of Energy, they could not manage to get down the 2.3 miles to the source of heat. They gave up, packed up their drilling rig, and silently stole away.

Drilling for heat is a much less scientific pursuit than drilling for oil. Oil geologists pretty much know where oil is, and, until recently, oil has usually been found at reasonable depths. For the most part the drillers also know what kind of oil they will reap—this from decades of experience and mapping of potential sites. The problem with drilling for heat is that one never knows what pollutants will come up with the heat.

Whatever does come up, be it hot, super saltwater or steam, bears with it a pharmacopeia of nastiness. This geothermal brine has been described as being some of the skankiest water to be found anywhere. Either it must be pumped back down under pressure, thus decreasing the net advantage in energy of the geothermal plant, or it must be made potable for release into our water systems. Too much of this noxious brew, because of sporadic regulation, already goes directly into our rivers and streams.

Unclean air in addition to unclean water is also a by-product of the geothermal plants now in operation. According to an article in *Energy Advocate* in 2001, "Somewhat surprisingly, the hot air that carries the heat from the earth at geothermal sites also contains carbon dioxide, CO_2, the Greenhouse Gas that sends anti-technologists into a frenzy. The suite of 20 geothermal sites from which Southern California Electric is obligated to buy electricity produces 730 tons of CO_2 for every gigawatt-hour (GWh) of electricity produced." More startling still, "Across the spectrum of all natural-gas power plants in the U.S., the average CO_2 production is 519 tons per GWh. That is, the geothermal power plants, which burn no fossil fuels whatsoever, produce 41% more CO_2 than the run-of-the-mill natural gas power plant for the same amount of electrical energy produced."

That geothermal energy produces almost half again as much CO_2 as a natural gas-fired plant does not sound like progress. These emissions approach the level of the CO_2 output of generating plants burning oil, and they are not too far below the output of plants burning coal, the worst of the greenhouse gas emitters.[29]

In addition to CO_2, other egregious gases come up with geothermal heat. Nitrous oxide, hydrogen sulfide, sulfur dioxide, and particulate matter are all a lesser part of the damaging brew. And though geothermal is always presented as a baseline source of energy, because the heat is always there, figures from the industry itself indicate that these plants must be taken down for a month or two each year for maintenance.

Central to use of geothermal energy, is the fact that access, primarily closeness to the surface, is widely and thinly distributed. There is plenty of heat, but in most parts of the world the cost required to get to it is prohibitive, and those areas where heat is retrievable might be, as with wind power, sited far from population centers.

Taking into account greenhouse gas emissions, siting problems, legal confrontations, difficulties and uncertainties in drilling, water pollution, earthquakes, downtime, and other eventualities, geothermal energy is far down on the scale of what is best for our industrial society and for the planet itself.

Hydroelectric

The three favorites of the green energy revolutionaries are hydroelectric power, solar, and wind power. All have been given a pass as emitters of greenhouse gases, but hydroelectric power has the advantage over wind and all of the others in the public's commonsense understanding of how the process works. What could be more clear and innocent than the weight of water turning a wheel? That image has been etched into the minds of generations of farm kids as they watched small streams push the waterwheels around that milled wheat between two heavy stones.

In modern hydroelectric generation, energy is plucked from the water that falls as rain cost-free from the sky. By the construction of huge dams, the water from the sky is prevented from running downhill and is held at altitude in artificial lakes until it is called upon to turn the turbines at the base of the dams. The free-running streams that pushed the mill wheels of the past depended upon the energy of the moving water. In modern hydroelectric plants it is the weight of the column of retained water pushing against turning blades that sends water screaming through the turbines at Mach-like speeds. As we shall see, the weight of the water that creates the energy to turn the turbines becomes, paradoxically, the very reason hydroelectric power generation as a cure for global warming may turn out to be no cure at all.

It seems a perfect and ideal process but, as always when humans endeavor to alter natural cycles, there are unintended consequences. One possibility is that the natural water cycle, upon which hydroelectric power generation is built, could fail. This failure could be the result of long-term natural planetary climate events. Or it could be a distortion traced to the recent impact of the changed relationship between carbon sequestered in the atmosphere and carbon sequestered in the sea and in land and forests.

According to a pair of researchers at Scripps Institution of Oceanography, UC–San Diego, "There is a 50 percent chance Lake Mead, a key source of water for millions of people in the southwestern United States, will be dry by 2021 if climate changes as expected and future water usage is not curtailed."

When I visited Hoover Dam in 2008, I was shocked to see how far down the intake columns the water level of Lake Mead had fallen. The all-time high-water level for the lake of 1,229 feet was reached in 1985. By the year 2000 it had fallen slightly to 1,215 feet, and then, in only the next four years, the bottom dropped out and by October of 2010 it had dropped to 1,080 feet. If and when the prediction that Lake Mead would disappear by 2021 holds, we will be left with the most expensive piece of concrete sculpture ever fashioned.

According to Hoover Dam engineers, if the water level falls much more, the dam will cease to produce power and shortly after that it will cease to be a source of fresh water for millions of folk in the deserts of our West. The water level in Lake Meade in 2011 stood at 1,087 feet, 65 feet below its level only five years prior. In those five years electrical generation had fallen from its rated capacity of 2,080 MWe to 1,700 MWe. It will cease if the water level falls only another 44 feet.

The gloomy future of Hoover Dam is not the future of all hydroelectric dams but, with weather patterns becoming ever more unsettled due to desertification and other global warming events, the economics of building more hydroelectric capacity is quickly becoming too risky, since we cannot accurately predict further local changes in the water cycle.

This calls into question the three positive claims that are made for hydroelectric generation. They are low cost, dependability, and longevity. If the useful horizon of a hydroelectric dam is reduced by one order of magnitude, that is, from hundreds of years to tens of years, it is clear that the economic value of existing dams will be sharply reduced, as would be the value of new construction.

In the search for renewable sources of energy, especially in the case of hydroelectric generation, we cannot assume that the future will replicate the past. It is quite possible that due to weather-generated water shortages, the thinning out of new sites for hydroelectric installations, and the high risks attendant on new construction we will never again see an increase in hydro-powered energy. In fact, if the sad tale of the Hoover Dam is any indication, it is likely that we have long since reached a worldwide peak of hydroelectric energy and can logically anticipate a slow but steady decline.

One other claim made for hydroelectric production is that hydro power is essentially clean power as compared to all other fossil fuel–burning installations. Proponents of hydroelectric power insist that there is zero emission of CO_2 attributable to electricity produced by free gravity and "falling water." The great dams that back up their great lakes directly convert the weight of the water above the dam into electrical energy, absent any use of CO_2-producing fuels.

To this point it would seem that hydroelectric power was the perfect, if limited, answer to the emission of global warming gases. But on deeper investigation it turns out that hydro power indirectly produces as much CO_2 as even coal-firing plants, and perhaps more than some.

"The green image of hydro power as a benign alternative to fossil fuels is false," says Éric Duchemin, a consultant for the Intergovernmental Panel on Climate Change. According to Philip Fearnside, of Brazil's National Institute for Research in the Amazon, in Manaus, "Everyone thinks hydro is very clean, but this is not the case. Hydroelectric dams produce significant amounts of carbon dioxide and methane, and in some cases produce more of these greenhouse gases than power plants running on fossil fuels. Carbon emissions vary from dam to dam."

In a study to be published in *Mitigation and Adaptation Strategies for Global Change*, Fearnside estimates that in 1990 the greenhouse effect of emissions from the Curuá-Una dam in Pará, Brazil, was more than three and a half times what would have been produced by generating the same amount of electricity from oil.

He goes on to explain that when a reservoir is created, it submerges enormous tracts of trees and plants, which settle to the bottom and begin to decay. Decay, or rotting, is essentially a burning process in which the carbon content of the bio material at the bottom of a lake produces CO_2, which is dissolved under pressure in the lake. A further and more serious problem is that this reservoir, with its cold bottom layer into which new bio materials are drawn, is low in oxygen, so that the rotting often produces CH_4, methane, and nitrous oxide rather than CO_2.

The initial decay of submerged material is then constantly replenished by organic materials brought down from the rivers and streams that feed the lake.

After the initial decay created by the backing up of reservoir water over existing flora, the process is continued, for the very reason that the dam was built in the first place. Dams are built to save water during wet times and release it during dry times. This "draw-down" creates a cycle of increasing and decreasing seasonal levels in reservoirs, resulting in the growth of flora on verges uncovered by ebbing levels in a dry season and then engulfed when the wet season appears. This injects a cyclical and inexhaustible supply of carbon into the reservoir.[30]

In a free-flowing stream, water is drawn off from the low-pressure top layer, which is richer in CO_2, rather than the lower levels, which are richer in methane. However, in a reservoir, in order to take highest advantage of the weight of the "head" of water to drive turbines, water is drawn from the lowest level, which is under high pressure and which has sequestered the highest levels of both methane and CO_2.

When these waters reach the turbine blades, the gases are subjected to the roiling fluxes and ebullition of water at the outflow and the sequestered greenhouse gases are blasted out of the solution. Philip Fearnside, the climatologist who first blew the whistle on claims that hydroelectric power was emission-free, liked to compare what happens at the bottom of a reservoir, from whence the turbines take their power, to opening a bottle of Coca-Cola. Bubbles of CO_2, previously held in solution by the capped bottle, now blow out of the bottle. If enough Coke is released into a confined space, the bubbles will kill you.

Indeed just such an event took place in Africa at Lake Nyos in Cameroon in 1986. The lake suffered what is called a limnic eruption. This occurs when carbon dioxide (CO_2) from the bottom of very deep lakes is suddenly released in a tsunamis-like event. The very cold bottom water becomes heavily saturated with CO_2 until the weight of the warmer water above can no longer contain the pressure created by the supersaturation. The CO_2 released in the Lake Nyos eruption instantly killed all living things, people, cattle, birds—all animals—for fifteen miles around.

A study by National Sun Yat Sen University and the National Central University of Taiwan supported the contention that the methane released is 20 times more harmful to the atmosphere than CO_2, and that nitrous oxide is an astounding 200 times more deleterious. We can no longer consider hydroelectric power a clean replacement for biofuel-burning electrical generation.

There is one other risk that emerges from the creation of huge bodies of water over what had been dry land, which is the possibility that the increased weight of water concentrated on a small area might cause seismic activity. The best test of this proposition is the effect of the Three Gorges Dam, which spans the Yangtze River. Seismic activity had long been recorded before the dam was built. This is now being compared to the level of activity postconstruction, and the results are worthy of note, as the reservoir sits on two major faults: the Jiuwanxi and the Zigui-Badong. According to Fan Xiao, a geologist at the Bureau of Geological Exploration and Exploitation of Mineral Resources in Sichuan province, "When you alter the fault line's mechanical state it can cause fault activity to intensify and induce earthquakes." By mechanical state Xiao means simply the additional weight of the water being loaded onto the area behind the Three Gorges Dam.

In 2003, when the filling process began, Xu Guangbin, director of the Hubei Seismological Monitoring and Prevention Centre, reported that "as many as 1,000 micro-earthquakes have occurred in the Three

Gorges reservoir area since June 7, with the biggest recorded at 2.1 on the Richter scale." These caused little damage or concern until, in 2006, when the reservoir reached 156 meters above sea level, the strongest earthquake to hit Hubei province in two decades shook an area near the Three Gorges Dam, reaching a magnitude 4.7. The tremor damaged thousands of houses and forced almost 6,000 people to leave their homes. The quake was centered in Suizhou City's Sanligang township, less than one hundred miles northeast of the Three Gorges project, and it rocked buildings in Yichang City, near the dam.

For the seven months after this level of the reservoir was reached, 822 tremors were recorded. They were small, but seismologists in the area remain concerned. The tremors will be of special concern when the level reaches its full 175-meter height.

Scientists Li Ping and Li Yuanjun, both civil engineering professors at Wuhan University, wrote, "A medium or strong earthquake would set off a chain of events in the reservoir area, with a series of landslides and riverbank collapses being triggered near the epicenter. The consequences could be dreadful to contemplate, quite unimaginable."

When the problems of water shortage, site shortages, and greenhouse gas production (especially production of methane and nitrous oxide) are taken into consideration, as well as the possibility of disastrous earthquakes, any dependence on the future of hydroelectric energy seems a foolish bet. The huge amount of capital that would be invested in new dam construction would be better used in almost any other energy project. Indeed, the time may soon come when the dismantling

of existing dams could end up constituting genuine, eco-
logically correct conservation.

Inanities and Insanities

Among those who endlessly churn out new schemes for
producing energy, there are those weirdly nice folk who
feel that they must contribute their ditzy bits to the
global warming scene. Some proposals are technically
interesting, some are hopeless, and some are simply
strange. Three Israelis were granted a patent for Active
Hanukkah Candelabrum, wherein the heat from tiny
Hanukkah candles is captured by a "sense circuit,"
which is converted into electricity. The energy imparted
to the dreidel by spinning it is also fed into the circuit.
Surprisingly, there are eleven other U.S. patent docu-
ments that are somehow related to this.

Considering that there are only a few million Jews
in the world and not all of them observe Hanukkah,
we can safely ignore the CO_2 produced by the burning
candles.

Another lunacy with a puerile twist is a small
tube suggested by the Bakuu Laboratories in which are
aligned magnets. Shaking the tube up and down vigor-
ously produces about 200 milliwatts. One observer
remarked, "Used as a penile insert you could probably
make your morning toast."

Still another has the dual intent to both produce
clean electricity and deter drivers of speeding cars. A
team of scientists has created a prototype of a speed
bump that can capture kinetic energy from vehicles
passing by, converting it into clean green electricity. The

prototype, called Motion Power Energy Harvester, is now being tested at a Burger King in New Jersey. To demonstrate how serious these folks are, engineer Jerry Lynch declared, "If this is multiplied by ten times the length and we have 100,000 or 150,000 cars a year, the device will pay back in less than two years." Of course, they do not acknowledge the warming input of the gasoline being burned or the cost of replacing shock absorbers.

Patent #4245640, issued by the U.S. Patent Office, is for a self-contained chest motion electricity-generating device that converts normal respiratory breathing motion into electric energy, which feeds into a magnetic circuit. The magnetic circuit is mounted on a chest harness that includes shoulder straps and a chest strap. The chest strap is made of an elastic material that expands and contracts with normal respiratory motion. That the respiratory motion is responsible for the emission of CO_2 is not factored into the saving.

Other human-powered devices are found on a site called Pure Energy Systems Wiki. They include:

- Energy-Generating Spinning Wheel Provides Power for the Poor: A charkha, or spinning wheel, is a ubiquitous machine in India. The *e-charkha*, designed by RS Hiremath, not only produces yarn, but also generates electricity and stores it using a maintenance-free lead-acid battery. (*Inhabitat*; Oct. 27, 2008)

- The Chukka Kinetic Music Player combines electromagnetic induction with a unique design that encourages the user to twirl it around the fingers, throw it about and otherwise toy with it. The result—an eco-friendly personal media player that also gives you the stress relieving benefits of tactile interaction and repetitive physical motion (*GizMag*; August 11, 2009)

- The nPower PEG (Personal Energy Generator) is a device that harvests the kinetic energy of your walking and turns it into electricity, charging most devices up to 80% with just one hour's worth of walking.

- Strike-Heel Generation: This is a piezo-electric heel-strike generation systems that derives power from the "good vibrations" of your heel striking the ground.

- Backpack straps harvest energy to power electronics: All that rubbing of your backpack straps on your shoulders may be put to good use, now that researchers have designed a novel type of energy harvesting backpack with straps made of a piezoelectric material that can convert the mechanical strain on the straps into electrical energy.

- Generating Power from Revolving Doors: *Fluxxlab* in New York is developing a "Revolution Door" that is basically a turbine powered by people as they enter or exit a building, using a central core to convert the motion of the spinning door into electricity.

- Yo-Yo Powered MP3 Player: The inventors estimate that between 10–12 tosses per hour are sufficient for continuous music play. A wireless headset allows the user to listen to music while yo-yoing.

- Lamp Lit by Gravity: A Virginia Tech student has created a floor LED lamp that is powered by gravity, using a weight slide similar to the mechanism that powers a grandfather clock. The lamp was expected to last two hundred years and was originally calculated to put out the equivalent of a 40 watt bulb. Mysteriously, the report goes on to state that "these calculations were in error."

- Crowd Farm to collect energy: A couple of MIT students would like to harness the mechanical power of large groups of people. A responsive subflooring system made up of blocks that depress slightly under the force of human steps would be installed beneath a thoroughfare. The slippage of the blocks against one another as people walked would generate heat.

My all-time favorite announcement is this, about which I would rather not know more:

- Asian Child Mimics Hydro Plant: Energy savant child, Kazuo Hydrohuko, has discovered how to harness the power of municipal water fountains. The method doubles as an effective enema.

There is one intriguing suggestion that merits reproducing the press release from its developer, Radio Corporation of America. RCA claims that the "Airnergy Charger scavenges stray WiFi signals and converts them to DC battery power that you can use to charge your cell phone, music player, or other electronic devices." RCA calls the Airnergy a WiFi Hotspot Energy Harvester. The device is about the size of a cell phone, with a micro USB connector hanging off it. Inside is an antenna to receive 2.4GHz (802.11) WiFi signals, and a converter that turns the broadcast energy into DC power, which is then stored in the on-board battery. You can keep the Airnergy in your briefcase or your pocket, and whenever it is within range of WiFi it charges itself.

RCA says that its researchers are already working on a new version that will be small enough to fit inside a cell phone battery. With the Airnergy harvesting battery pack, you could recharge your phone or other device simply by leaving it range of a WiFi hotspot. All those multifrequency waves in our world must be good for something. Considering that direct recapture of

CO_2 might not work, maybe this can save a little greenhouse gas.

When how much things cost is not a concern, as in all military decisions, the U.S. Navy's reach for energy from the sea turns silly. A process that the navy is actively pursuing has a price tag that would preclude any development save by the bottomless pockets of the Department of Defense. The navy is building an Ocean Thermal Energy Conversion (OTEC) plant the size of an oil platform that would require a pipe 33 *feet* in diameter and 3,000 feet long.[31] This monster pipe is designed to pump cold waters rich in sequestered ammonia from great depths and to precipitate gaseous ammonia by using the warmer waters drawn up from shallower depths by another tremendously big pipe. This billion-dollar plot against the U.S. Treasury is predicted to *maybe* produce less than 100 kW, if it works at all. And should it actually produce something, a future admiral, contemplating the manning, operating, and maintenance cost of the floating skyscraper-sized platform on which it is mounted, will junk it and simply go out and buy a small diesel generator for about a millionth of the cost of construction and a billionth of the cost of operation.

By any standards, this project seems to be the most inane of all and, ironically, the one that will most likely be tested with tax dollars. All of the others have the enormous advantage of being unable to tap into the public pocketbook. These will live and die in obscurity, but the U.S. Navy OTEC project, like other military fantasies, such as a nuclear-powered airplane, will be well funded and, in the end, relegated to an enormous warehouse, along with other military projects that should never have been allowed to see the light of day.

One can but wonder where they will be able to stash a pipe 33 feet across and more than half a mile long.

Meat and Milk

If an astounding 12 percent of worldwide greenhouse gases are due to the production and consumption of meat, as was reported by the Food and Agriculture Organization (FAO) of the United Nations in 2006, then we had best sit up and pay close attention. This estimate equates our penchant for meat with our dependence on fossil fuels, road transport, and concrete. The FAO report estimated that, as of the turn of the millennium, meat animals were pouring 7,500 million tons of CO_2 equivalents into the atmosphere every year. That level of pollution, let alone the increases predicted, is clearly insupportable if we are to address the warming problems that we face.

But then two respected researchers took a look at the methods and results of the FAO and discovered four areas that had been uncounted or misallocated in the study. Robert Goodland and Jeff Anhang of the Worldwatch Institute now suggest that, when statistical oversights are taken into account, the total impact of meat animals is, in fact, a shattering 51 percent of all warming gases emitted throughout the world.

It may be that the FAO's 12 percent was an understatement and Goodland and Anhang's 51 percent was an overstatement. Nevertheless, any percentage between these two extremes represents a clear and present danger that must be considered and quickly dealt with. David Steele, of the University of British Columbia, arrived at a figure of 30 percent. Whether the correct figure is 12 or 30 or 50 percent, the world gluttony for meat is a searing problem for the planet.

World livestock population has curved upward along with world population and has been, and will be, magnified even further by the new populations of potential meat eaters emerging from formerly third world nations. The Chinese population alone will represent an additional burden on the ecology as the Chinese people move from rice to meat as a staple of their diet. The change in China is made possible by an increase in national wealth. China has doubled its consumption of meat in just a few decades and is on the verge of another doubling as its peasantry emerges into a growing middle class.

The same change may or may not be possible in the short term for the rest of the third world, but they remain in the wings as a probable additional meat-eating threat.

The gluttony for meat is matched by the gluttony of the livestock industry for our forests, which are being cleared for grazing.[32] Not only does the cultivation of livestock affect global warming, but it is among the top contributors to a whole range of environmental problems, including, in addition to global warming pollutants, land degradation, water shortage, and, as forests disappear, severe curtailment of biodiversity. As Steinfeld et al. note, "The total area occupied by grazing is equivalent to 26 percent of the ice-free terrestrial surface of the planet. In addition, the total area dedicated to feedcrop production amounts to 33 percent of total arable land. In all, livestock production accounts for 70 percent of all agricultural land and 30 percent of the land surface of the planet."[33]

Any consumption behavior pattern that uses up three-quarters of the world's agricultural land calls loudly for mitigation on a worldwide level. However, with most of the world already at subsistence levels and that portion of the population on the rise, it is difficult to see how behavior relating to eating, the most fundamental of life's drives, can be reversed within a meaningful time span. In highly organized third world economies, such as China's, behavior is being intensified in a direction that bodes ill for the way of life in the developed industrial world. Hunger in poor countries does not respond to what might happen in a century or in a decade or, in extreme cases, even tomorrow.[34]

The use of meat animals as a primary source of food is not only an attack on the environment, it simply does not make sense economically. Almost any nutrient material, grains, seeds, beans, nuts, or leaves, will feed more people, more healthfully than an equivalent amount of meat. The food chain that flows from sun-nurtured flora to flesh-eating fauna is an extremely inefficient process. It has been demonstrated that a large percentage—often estimated at 90 percent—of energy is lost in the conversion of wheat to meat. This means that it would take more than ten times less of the amount of the grain now dedicated to meat production than if the eaters were to consume the plants directly. It also follows from this argument that it takes ten times more land, ten times more fertilizer, and ten times more fuel in the process of creating meat. Separate studies, which vary widely but whose estimates are egregious at either end of the scale, suggest that producing meat takes from 30 to 300 times as much water, our most scarce resource, as it would to produce an energy-equivalent amount of grain.

There are problems related to livestock other than CO_2. Animal manure generates nitrous oxide, a greenhouse gas that has almost 300 times the warming effect of CO_2. Also, the emission of gases from both ends of cows produces methane, which is, as has been noted, also much worse for the atmosphere than CO_2. Each cow produces up to 200 pounds of methane gas each day. The United States alone has 100 million cattle, contributing daily to the degradation of the atmosphere.

Noam Mohr, of EarthSave, makes a convincing case.

> By far the most important non-CO_2 greenhouse gas is methane, and the number one source of methane worldwide is animal agriculture. Methane is responsible for nearly as much global warming as all other non-CO_2 greenhouse gases put together. Methane is 21 times more powerful a greenhouse gas than CO_2. While atmospheric concentrations of CO_2 have risen by about 31% since pre-industrial times, methane concentrations have more than doubled. Whereas human sources of CO_2 amount to just 3% of natural emissions, human sources produce one and a half times as much methane as all natural sources. In fact, the effect of our methane emissions may be compounded as methane-induced warming in turn stimulates microbial decay of organic matter in wetlands—the primary natural source of methane. Methane is produced by a number of sources, including coal mining and landfills —but the number one source worldwide is animal agriculture. Animal agriculture produces more than 100 million tons of methane a year. And this source is on the rise: global meat consumption has increased fivefold in the past fifty years, and shows little sign of

abating. About 85% of this methane is produced in
the digestive processes of livestock, and while a single
cow releases a relatively small amount of methane, the
collective effect on the environment of the hundreds
of millions of livestock animals worldwide is enormous.
An additional 15% of animal agricultural methane
emissions are released from the massive lagoons used
to store untreated farm animal waste, and already
a target of environmentalists for their role as the
number one source of water pollution in the U.S. The
conclusion is simple: arguably the best way to reduce
global warming in our lifetimes is to reduce or eliminate our consumption of animal products.

In the light of all the waste and spoilage and damage to the ecology that meat eating involves—and, not
incidentally, in light of the damage it does to our own
blood vessels and hearts—it would seem that the only
solution to a problem that is growing exponentially is a
universal behavioral shift, in both the rich and the poor
parts of the world, away from the eating of meat. The
benefits of such a movement are obvious. Unfortunately,
it is highly unlikely that any change will occur.

Geophysicists Gidon Eshel and Pamela Martin
estimate that, if Americans would eat only four meat
meals a week instead of five, it would be like trading in
all of our gas-guzzling clunkers for hybrid Priuses.

There is another way to think about all this meat
hunger. If Americans would desist from the eating of
only hamburgers, 200 million fewer metric tons of CO_2
would be emitted each year—the equivalent of taking
40 million cars off of America's roads.

And if these were not sufficiently scary statistics,
a United Nations report in 2006 found that the meat

industry produces more greenhouse gases than all the SUVs, cars, trucks, planes, and ships in the world combined.

Natural Gas

If only because it produces half the amount of CO_2 as coal, natural gas lays a more gentle hand on warming trends.

Having only one carbon atom in its makeup, natural gas is the cleanest-burning of all the carbon-based fuels. Unlike wind and solar power, which cannot be stored and which require expensive backup of fossil fuel–fired plants, natural gas can be stored either in its gaseous state or, as liquefied natural gas, in a low-temperature liquid state.

But all of this requires an expensive infrastructure. Retrieving natural gas from underground itself is not a costly procedure—in some cases it is simply a by-product of retrieving oil—but once it comes out of the ground it must be immediately piped away or stored. Natural gas is methane, a highly corrosive, highly explosive, and highly lethal gas.[35] In order to control these characteristics, natural gas storage tanks require higher amounts of weight and space than do oil. The strict storage requirements, the high maintenance costs in storage, and the risks and the cost of transporting natural gas add so much to the price to the user that natural gas eventually becomes competitive for only the highest-value uses.

Because of its storability, the energy in natural gas is capable of fast ramp-up in the case of loss of generation capacity elsewhere. However, this use of natural gas as a reserve not only puts CO_2 in the atmosphere due to the energy needed to pump and compress it, but it also adds methane from inevitable leakages.

A serious problem emerges from the uneven distribution of natural gas reserves. Over 50 percent of the usable reserves are located in Russia, with another 15 percent in Iran and 14 percent in Qatar, countries whose long-term aims are hardly consonant with the interests of the industrial West. This represents the threat of oligopolistic control of energy and a serious geopolitical problem if producing countries are tempted to use access to natural gas as a weapon in international affairs, as occurred recently between Russia (the producer) and Ukraine (the user).

This oligopoly, similar to the cartel that controls the flow of oil, is reason enough for industrial nations to actively work to limit the import of natural gas. Significant wealth is exported for our use of oil. That amount should not be added to by the importation of this even more expensive energy source.

A counter to the foreign control of natural gas fields is the recent development of fracking in gas fields in the United States. Fracking is best understood through the words of one of the chief frackers in the natural gas industry:

> We drill down to the target zone; we run a steel pipe beside the hole and cement it in place, then we put perforations in that pipe in the area that we want to produce the oil or gas from, and in the case of a fracture-stimulated well we actually will go in and pump fluid under high pressure into the formation and that effectively cracks the rock...and we pump in some sand...that will hold those cracks open. And what those cracks become are the flow-channels through which the oil and gas can reach the well bore.[36]

What this description leaves out is that the "fluid" that is pumped down to break open the rock structures is billions of gallons of increasingly scarce water. Furthermore, what is also not often mentioned is that the energy legislation of 2005, crafted as a result of former Vice President Dick Cheney's secretive energy task force, exempted fracking from the Evironmental Protection Agency's Safe Drinking Water Act and left regulation in the hands of individual states, with all of the attendant opportunities for evading regulation at the state level.

This is not a marginal consideration. Since 2008, when huge reserves of natural gas began to be extracted from the Marcellus Shale under Pennsylvania and New York, 4,000 gas wells have been drilled in Pennsylvania and 14,000 in New York State. In 2009, it was revealed that the New York Department of Environmental Conservation had insufficient staff to monitor water pollution. In 2010, the state of Pennsylvania still did not have a serious regulatory mechanism in place, nor had the state legislature been able to agree on any level of taxation on the extraction of natural gas. At this writing, regulating the degradation of water due to fracking is in complete disarray.

The foreseen and unforeseen problems that involve the degradation of drinking water by fracking have yet to be taken into account in the real costs of fracked natural gas. When the water used to rip apart the gas-bearing rock is returned to the surface, it is laden with a soup of any of two hundred or so chemicals, some of which are known carcinogens. This water is disposed of in streams and rivers with little testing or control. Some of the fracking water that is not pumped back out but remains puddled amid the fracked rocks has been reported to spread as much as 30 miles underground from the wellheads, with consequences that are fast becoming threats to local water supplies.

This February 2010 report from Harrisburg, Pennsylvania, published in the *Malaysia Star*, demonstrates some of the serious real-life problems that are emerging from fracking.

Fracking a horizontal well costs more money and uses more water, but it produces more natural gas from shale than a traditional vertical well.

Once the rock is fractured, some of the water—estimates range from 15 to 20 percent—comes back up the well. When it does, it can be five times saltier than seawater and laden with dissolved solids such as sulfates and chlorides, which conventional sewage and drinking water treatment plants aren't equipped to remove. At first, many drilling companies hauled away the wastewater in tanker trucks to sewage treatment plants that processed the water and discharged it into rivers—the same rivers from which water utilities then drew drinking water.

But in October 2008, something happened that stunned environmental regulators: The levels of dissolved solids spiked above government standards in southwestern Pennsylvania's Monongahela River, a source of drinking water for more than 700,000 people.

Regulators said the brine posed no serious threat to human health. But the area's tap water carried an unpleasant gritty or earthy taste and smell and left a white film on dishes. And industrial users noticed corrosive deposits on valuable machinery.

One 11-year-old suburban Pittsburgh boy with an allergy to sulfates, Jay Miller, developed hives that itched for two weeks until his mother learned about the Monongahela's pollution and switched him to bottled or filtered water.

There are urgent reasons to resist the increasing use of natural gas. The first is that it is a source of CO_2 and methane, which will have to be dealt with at enormous cost in the future. Second, as more and more "natural" sources of gas are depleted, the industry is turning to fracking, with its enormous demand on scarce water supplies. Third, we still do not have even the vaguest idea of what the unanticipated consequences will be of ripping apart rock structures that underlay heavily populated areas.

While natural gas is a "cleaner" fuel than coal, its use still adds millions of tons of warming gases. Reducing CO_2 output by burning natural gas rather than coal does very little other than delaying for a short while the worst effects of global warming. The only rational way to mitigate now the threats to the future is to ban all CO_2-producing fuels, the "better" along with the worse.

Oil

The trouble with oil from the conservationist point of view is that, like coal, there is too much of it, it is too easy to suck out of the earth, it is a very good heat source, and it is cheap.

On the other hand, the only concerns of the energy oligarchs are that the sources of oil are specific and local, and it must be transported expensively over inconvenient distances before it can be blown out of the tailpipes of hundreds of millions of internal combustion engines.[37] Within the present levels of price and demand, this is not a substantial problem for the industry. In fact, there is nothing in the production or consumption of oil that represents a threat to the existence of the oil industry. Quite the opposite: everything about oil predicts that it will continue to be profitable.

For the conservationist, the fundamental trouble with oil is its long-term existential cost. Oil's availability and thermal content define it as a leading player in the warming of the Earth as well as the warming of our hearths. Estimates of the amount of reserves exceed 15 trillion barrels of retrievable oil. At its present rate of consumption of 30 billion barrels per year, it would take 450 years to suck up the last barrel.[38] This is not an acceptable condition in light of the looming warming trend. The fact that billions of barrels of oil will be available each year for as far into the future as we can conjure poses life-and-death choices for the oil industry and for the rest of us, and for the climate to which we have become accustomed.

Oil is responsible for billions of tons of CO_2 emissions,[39] equaling the emissions of coal, even taking into account that the thermal content of oil is less polluting than that of coal. This tonnage represents about 10 percent of the total world emissions, a substantial fraction. But this fraction in no way reflects oil's economic stranglehold on our economy and on our general short-term well-being. While there are immediate substitutes for other energy sources, should any one be withheld, there is no substitute for the unique function of oil as the mover of transport in its present form. Within a few days of a constriction of the flow of oil, as was demonstrated by the 1973 OPEC oil embargo, the industrialized West can be brought thirstily to its knees.

Significant change in the consumption of oil might be achieved gradually in the behavior of the millions who transport themselves and their goods across the roadways of the world. The difficulty in changing behavior has been clearly shown in the matter of idling engines. Idling an internal combustion engine uses more fuel than shutting down and starting up. This simple and obvious saving to both the individual driver and to society should be easy to apply. Yet it has been almost impossible for government to make a dent in the billions of traffic light stops that occur each day in spite of the fact that any time an internal combustion engine is allowed to idle for more than 10 seconds it is cheaper, in terms of fuel consumed, to turn the engine off. This suggests that any decrease in the use of oil as fuel will have to come from above, by governmental decree, and that mass behavioral changes are unlikely or will appear too late to affect warming trends.

The oil industry's nonreplaceability creates an analogue of monopoly, and monopoly creates excess profits and unassailable economic power. As a result, the oil industry seems immune to a global warming crisis. Indeed, as the planet warms, we survivors may well need ever increasing reserves of industrial energy such as oil to counter warming symptoms and events. Thus the rise in the influence of the oil oligarchs over our society, and their intransigence, seems inevitable.

The future looks equally bleak if one looks at the demand patterns as the standard of living in the under-consuming world improves. The first desire on entering the middle class is for personal transportation: a car. The United States has 250 million registered vehicles in a population of 300 million. China has approximately 25 million vehicles in a population of 1.4 billion. A reasonable prediction, considering the industrial growth rate of China, would be that China will have 500 million vehicles within a decade.

Considering this, it is hard to imagine any program that might cause the CO_2 emission rates for oil to hold steady, let alone decrease. And yet as we look at the climbing parts per million of CO_2 in our atmosphere, we know that changes in oil consumption must be made and will be made. The question remains whether the changes made will be too late.

Since we are not likely to excise the yearning for personal transportation from a growing world population, and since we are not likely to successfully challenge the entrenched powers of the auto and oil industries, we must look for another model for individual mobility.

A possibility, in fact the only possibility that might create a transportation industry without destroying much of our present industrial capital or precipitate an economic civil war, will be the development, already under way, of electric vehicles powered by nuclear sources. The auto industry has already joined the trend, and the oil industry, facing a move from liquid fuel to electricity, will be forced at some point to move its enormous resources from the extraction and refining of oil to the production of nuclear energy. Few industrial complexes are rich enough to contemplate such a move. The oil industry, in its own self-interest, has the economic ability and the engineering smarts to use nuclear as its future in transportation. But engineering smarts are not equivalent to long-term political smarts so again the question comes down to whether an entrenched and hugely profitable industry can overcome its internal resistance to change.

Electric cars bring their own problems, some as yet unforeseen, some we know we will have to confront. When you move from a few thousand electric cars to some millions, the effect of draw-down of power of a magnitude required to service millions of recharging vehicles simply can not be handled by our present grid, or by however smarter our grids can be made. What will be required will be limiting the reach of continental-sized grids and replacing them with regional small nuclear generation. Such plants, capable of serving tens of thousands of homes with little loss during transmission, will bring power to the people in packets that cannot precipitate national energy crises. Gigawatt power plants are, like our banks, too big to fail. And yet historical warnings were ignored, and the banks did, and the grids will too unless we look to an entirely different profile of energy distribution.

Solar

All that is required to extract more energy from power-generating systems is to increase the input of its fuel. In most systems the amount of fuel injected is a variable that is under our control. Dig more coal, extract more natural gas, build more dams, or mine more uranium and energy shortfalls disappear.

With solar energy, however, the amount of "fuel" is immutably given. The average amount of sunlight reaching the surface of the Earth is about 30 watts per square foot—or, factoring in the predicted optimistic 50 percent efficiency of future solar panels, 15 watts per square foot. This is, in the large picture of our energy needs, a trifling amount[40] that is further reduced since most energy demand is in the mid- and high latitudes, where most of the world's population resides and where there is the least sunlight per square foot.

At this rate, considering all of the limiting factors, and optimistically assuming that a full 50 percent of solar energy can be converted to electrical energy, we would need to dedicate six square miles of unused desert land, a resource that is absolutely finite, to replace just one coal-fired power plant. To make a significant contribution to our total energy requirement would require millions of square miles of scarce and sunny land, most of it far distant from users.

Things are a little better if we consider thermal collectors with mirrors to heat water to the temperature of usable steam. To provide the world's demand for energy by this method, we would still need to use 250,000 square miles of scarce hot desert. But to construct the couple of billion collectors that would be required would cost trillions. The constant maintenance required to keep billions of mirror surfaces and solar panels clean and efficient is almost incalculable. These are daunting numbers. Even if we are trying to reach a more modest goal of supplying only 10 percent of our needs by this method, the costs and the ecological confrontations of solar energy would severely test the will of the energy community.

Sunlight is free, but solar panels that generate electricity are not. For the average household, which pays about $120 a month for cheap coal-fired electricity, the installation of personal use solar panels would require years to break even. Even if the price of solar panels comes down as the industry develops, the break-even point for a household would still likely be in decades.

All of these objections pale when deeper investigation reveals that any use of solar-powered energy would supply a significant fraction of carbon to the atmosphere: a condition that solar is specifically touted to avoid.

The situation is identical to the objections to wind power, where we recognize that solar, like wind, is a discontinuous source of energy, which ceases as the sun goes down. When clouds cover the sun during daylight hours, electricity flow diminishes, and if the flow is any significant portion of the energy being supplied to the grid, immediate replacement of the missing energy is required. Replacement energy can come only from those forms of power generation that can be quickly ramped up, such as power plants that use fossil fuels. If we were to discontinue the use of all solar energy collection, and wind as well, there would be, due to the requirement for standby fossil fuel–generating systems, a net reduction in the total release of greenhouse gases.

One alternative backup model that some advocates of solar power are proposing is the use of molten salt to store solar energy. This scheme proposes an array of 20,000 mirrors, each measuring 24 x 28 feet, all focused on a 500-foot tower in which molten salt is heated to 1,000 degrees. The claim is that such a device would produce 150 megawatts (MWt) of power, which is about 35 megawatts (MWe) of electricity. There has been no estimate of the cost of construction and installation, but one firm, SolarReserve, claims that their research costs alone have reached $100 million.[41]

To keep matters in perspective, just one of the small nuclear plants (SMR) would replace this complicated system. A proposed Babcock & Wilcox mPower miniplant would cost, assuming the assembly-line methods proposed by the developers of SMRs, in the neighborhood of $100 million to $200 million. These modular units would produce the same amount of energy as the molten salt proposal, but maintenance costs would be orders of magnitude less.

Understanding that solar power brings with it an indirect emission of greenhouse gases is a clear refutation of the mantras of the Rocky Mountain Institute, which demands the end to all nuclear generation in favor of "renewable" sources such as wind and solar. If we were to follow the lead of nuclear deniers such as the RMI, the day of the tipping point, beyond which we will no longer be able to control planetary warming, will be brought much closer. The tipping point may already be perilously close to that fateful planetary capacity for rapid man-made changes, beyond which we will have no say at all about the warming of our planet.

Finally, there is the water problem. There is always a water problem. Little has been published about the enormous draw on water that large solar arrays represent, especially in the preferred arid desert areas in which solar plants are the most efficient and where water is most scarce. In some such plants, mirrors create steam that drives turbines that make electricity. Just as in a fossil fuel power plant, which solar advocates seek to replace, that steam must be condensed back to water and cooled for reuse or must be constantly replenished from scarce supplies.

The conventional method of reuse is called wet cooling. Hot water flows through a cooling tower where the water evaporates along with the heat. The lost water must be replenished constantly. One plant proposed in a dry area of California required 3 million gallons of water a day, which represented a significant percentage of the entire water supply for the region.

Dry cooling, an alternative to wet cooling, uses fans and heat exchangers, much like a car radiator. Far less water is consumed but use of electrically driven fans adds costs and reduces efficiency and would make dry cooling noncompetitive with other forms of generation, especially with small nuclear units, which do not require any vast acreage of sunny desert land and lay the lightest burden on ever more scarce water supplies.

There is no free lunch. Indeed, renewable dining turns into expensive banquets in unanticipated ways. In the case of harnessing the sun, in addition to a crucial water problem, there remains that component of CO_2 from backup systems, and finally there is the problem of siting in increasingly scarce land, the ultimate scarce resource.

Wind

"As free as the wind" is somewhat of an overstatement.

As long as wind-generated electricity produces less CO_2 than other systems, that fraction, and only that fraction of the CO_2 benefits, is entirely free.

But this positive view of wind refers only to the immediately direct process by which the pressure of the wind on the rotors results in electricity coming out of the generator. What happens after that raises awkward problems about the ultimate cost of making electricity from the wind.

The problems are:

Intermittence
Siting
Transmission loss
Bird collision
Aural and visual objections
Grid requirements

As any sailor knows, there are long days at sea when the wind does not blow at all, and there are longer days when the most that can be squeezed out of a desultory breeze might be a few nautical miles. When the wind does blow over the open sea, there are no structures that might make a breeze turbulent. It blows, when it blows, in a laminar flow, long and straight and without roil.

Since wind generators act best when presented with a laminar flow of air, the preferred locations for wind farms are far offshore, where turbulence from vertical structures and geographic changes in altitude are not a problem. Additionally, there are two other advantages of sea-farming the wind. There is no one around to be bothered by noise and, if carefully sited, the farms are far enough offshore not to be abusive to sensitive defenders of their pristine views.

Remembering that the world is round, the visual horizon is only three miles away, so if farms can be placed a reasonable five miles offshore, there should be, like the objections to noise, few visual quality-of-life confrontations. However, considering the needs for a high constancy of wind combined with ocean waters that are not too deep, there are relatively few areas where offshore wind farms could be sited.

When wind farms are brought ashore problems abound, of which site selection is the most important. There are precious few areas in most parts of the world where the wind is of sufficient force—say, an average of 15 mph—to make electrical generation financially feasible. Indeed, there are no sites—even at Cape Horn, at the southern tip of South America, which has the most fierce wind profile—where the wind does not sometimes just stop altogether.

High and consistent wind velocities are regularly found in areas far from urban centers. Whether located far at sea or in windy mountain passes, most wind generator installation will require expensive transmission lines to link up with the national grid. On land there will be, and already are, screams of dismay from landowners over whose property the unsightly towers will have to march. There are so many claims and players in this mix—including state and federal governments, wind farm developers, and competing systems already in place that use no wind—that it is hard to predict how this will sort itself out. Any loss of predictability raises the cost of risk to capital that could move the cost of wind energy beyond its ability to compete.

Wind, while seemingly free,[42] is fickle, and its capriciousness raises the question of whether wind-generated electricity, like solar, reduces the output of greenhouse gases. It might seem a stretch between the generation of wind energy and the output of CO_2, but the relationship is real and emerges from that most intransigent quality of wind energy, its intermittence. The experience of the Danes points up the delicate balance between how hard the wind blows, or does not blow, and the need to replace lost energy to a hungry grid. As little as a single meter per second drop in wind speed in Denmark results in the need for an additional 350 megawatts of electricity to be supplied from other sources. In fact, measured by the intensity of carbon that emerges from electricity production, Denmark, with its thousands of wind towers and its high level of wind consistency, still is nineteenth on a list of lowest carbon user nations and ten times the carbon intensity of France, which boasts the lowest carbon intensity of all nations.[43]

An extreme example[44] of the problem of intermittence emerged in New Zealand in 2006 when the wind stopped at the same time that some generating plants were taken down for maintenance. Auckland, New Zealand's largest city, had been paying an average of about $25 per MW hour. When the wind died that day, the price zoomed to $1,233 per MW hour.

National electric distribution grids are fed by all the many sorts of electric generation: nuclear, coal, oil, gas, hydro, geothermal, solar, and wind. Of these, coal, oil, and gas plants are designed to provide variable amounts of energy — as needed — for the grid. Solar and wind are intermittent and unreliable. When the sun goes behind a cloud, or sets, and the wind displays its inconstancy, their contribution to the grid must, as has been demonstrated, be instantly replaced. The problem is that there are no instantly up-rampable energy sources that are free of CO_2. Coal plants are not designed to quickly change output. Only oil and gas, if so designed, can quickly replace the loss from wind and solar. Thus when wind fails, it is replaced with CO_2-generating output, the absence of which was the best argument for wind power to begin with.

Building and maintaining oil and gas capability, used only a fraction of the time, is a waste of economic resources in addition to adding CO_2. One careful study[45] of the amount of CO_2 emitted by energy systems reliant on wind revealed some startling numbers. For wind power without backup, the emission of CO_2 is an inconsequential 18 kg of CO_2 per MW hour. This comes from materials and energy used to construct the turbines and towers. But when adequate backup from energy sources capable of ramping up to compensate for downtime of wind generation is factored in the emission rate rises to 519 kg of CO_2 per MW hour.

On August 19, 2009, in the *Duluth News Tribune*, reporter Rolf Westgard provided a graphic word picture of reliance on wind power. "Shortly after 6 p.m. on February 26, 2008, winds in West Texas died, and the massive blades on hundreds of wind turbines stopped. Within minutes, 1,700 megawatts of wind power supplying the Texas electric grid declined to 300 megawatts, threatening the integrity of the grid."

In any event the total present and predicted addition of energy to our present needs is far from significant. A report highlighting this came from the Electric Reliability Council of Texas. The council concluded, "Erratic Texas wind power will have an 8.7 percent capacity factor, or just 708 megawatts, providing 1.0 percent of the state's need for 2009. The Electric Reliability Council of Texas' forecast for 2015 shows wind energy rising to provide just 1.2 percent of the state's electric power."

Furthermore, the U.S. Energy Information Administration is forecasting that U.S. wind and solar combined will supply just 2.1 percent of the total U.S. electric power supply of 4,618 billion kilowatt-hours by

2020. This is far short of the goals of 20 to 25 percent of energy from wind envisioned by some enthusiasts.

To this observer, that is both bad and good news: bad because of the slight impact that wind has on our total needs, and good because we will not be living in a fool's paradise in which wind power, if indeed it does ever become significant, will continue to contribute its insupportable burden of carbon to our atmosphere.

Desirable wind, solar, and hydro sites are usually located far from the ultimate users. These "clean" energy systems are rarely sited relatively near the consumer, as coal and gas and nuclear generation can be. Since significant power is lost in transmission, the cost of building and maintaining long-distance grids must be factored into the cost of distant wind farms.

Other unintended obstacles include local opposition. Surprising resistance comes from groups that would be expected to support a "green" energy source.

The residents of Martha's Vineyard, generally and unrepentantly green, initially opposed an offshore wind farm in their bay because it impeded their view. It took ten long years to get approval for this well-sited wind farm. A slightly more reasonable objection is that flying creatures, such as bald eagles, falcons, and bats, find the rotors incomprehensibly attractive.

A major project initially involving $11 billion of new investment and eventually promising 5,500 MW of low-cost electricity is being held up by the unlovely but endangered lesser prairie chicken and by the more elegant spotted owls in Oregon. Until it can be guaranteed by developers that no further endangerment to these birds (the lesser prairie chicken has declined in numbers from a recent high of 3 million to the present approximately 10,000), delays and costs will mount. To save economically unimportant species from extinction for the sake of desirable diversity is a moral imperative that eventually will need to be balanced against the putatively higher claims of our own human species.

What is the bottom line in this particular matrix of ecological conflicts? When the added CO_2 and the siting problems and land use and transmission problems and the threat to wildlife and the passionate NIMBY (not in my backyard) objections associated with wind power are added up, it is difficult to look at wind power as free or as green as it is painted. In any case, it remains a stopgap measure on the road to a long-term solution to our growing energy shortages.

In the case of wind energy, the game might just not be worth the candlepower.

Sequestration

"The problem is that...CO_2 is only a trace component of the atmosphere, currently about 0.038% of the molecules in the air. It takes energy and work to unmix CO_2 from that dilute mixture. Releasing CO_2 to the atmosphere only to extract it back out again would be a really stupid energy strategy."

—Archer, *The Long Thaw*

A scenario.

The year is 2030.

Two decades earlier, in 2010, $2 billion was directed toward the sequestration of carbon emissions from the nation's biofuel-burning plants. In the year when the sequestration program was instituted, a little over 1 percent of carbon emissions was being sequestered. Backed by the coal, oil, natural gas, and mining companies, the sequestration program ballooned from 1 percent sequestration in 2010 to 90 percent sequestration in 2030. The cost of the program grew from $2 billion in 2010 to $2 trillion per year by 2030.

The cost of burying 90 percent of our carbon in the ground required a substantial part of our gross national product. Taxes skyrocketed, government services were cut, unemployment increased, the value of the dollar tanked, and, due to increases in population and increases in industrial activity, we were still putting 2 billion tons of CO_2 into the atmosphere each year. However, the process became unstoppable, as there were no excess resources available to develop long-term technologies that might have been used to halt rather than sequester carbon. The sequestration monster kept spiraling upwards with no end in sight.

However, 2031 represented the year that the entire sequestration effort came to a sudden halt, throwing millions out of work and causing the Great Depression of 2033, when it was realized that there simply was no place left in all of America to put the CO_2 that the sequestration plants were spewing out. The 18 billion tons of CO_2 emitted annually required underground storage volume of 30,000 cubic kilometers each year. Over the twenty years of sequestration, almost a million cubic kilometers of underground space had been filled. America was sitting on almost half a trillion tons of buried CO_2, with not a single additional square kilometer of storable underground space left.

With no place left to sequester another 18 billion tons of CO_2 annually, gas was dumped into the atmosphere, and the United States was back to 2010. Except that by 2035 the sequestered CO_2 was combining with underground water to form carbonic acid. The acid had started to dissolve the limestone caverns in which most of the CO_2 had been kept. This created great underground voids, the consequences of which were subsidence of the land and massive destruction of surface installations. To further complicate matters, the subsidence was accompanied by the invasion of carbonic acid into the already short supply of potable water.

The sequestration program in two decades had allowed the biofuel industry a furlough of twenty years in which to undermine the viability of the free economy of the United States. They left the nation and the world poor, polluted, and awash in CO_2.

LIKE ALL SCENARIOS of doom and gloom, this one omits all of the events that might have mitigated its outcome. But the huge quantities and volumes are correct.[46] If, as the scenario assumes, all other continuing processes were held in abeyance, the prediction describes exactly what would happen had we made this Faustian bargain that Alvin Weinberg[47] described in 2007. He and his team have labeled carbon capture and sequestration (CCS) as "little more than a free PR campaign" for the biofuel complex to divert attention from halting emissions by suggesting that emissions could be contained by sequestration.

They go on to say that any investment in CCS will extend the useful life of carbon-emitting plants, thus making economies even more dependent on biofuels. "Instead of buying us time to find alternate sources of clean energy, CCS is buying politicians' time to avoid making tough, unpopular decisions. The allure of CCS threatens to divert resources from energy efficiency and delay more durable reforms. CCS may be, politically, an easy way out of having to make more difficult and sustainable choices."

There are other unintended consequences of dealing with huge amounts of anything, no matter where they are placed. Phillip Boyd[48] identifies sequestration as a form of geoengineering and raises the concern that the "unintended detriments of geoengineering strategies such as carbon sequestration...might lead to conflict between nations." This is especially true of the impact that geoengineering schemes of the size required for sequestration would have on world water resources. He further argues that "it is the very scale and longevity of these schemes that makes regionally heterogeneous side effects more likely, and the potential for discord between nations more real. The unintended dispersal of geoengineering agents will only exacerbate the problem."

The fact is that we know very little about what can happen when large amounts of CO_2 are shunted about. A case in point is Lake Kivu, in Democratic Republic of the Congo, which now contains 300 cubic kilometers of carbon dioxide and 60 cubic kilometers of methane that have bubbled up from volcanic vents. The gases are trapped in layers 80 meters below the lake's surface, held there by water pressure. However, geological or volcanic events could disturb these waters and release the gases, with a devastating impact similar to the August 1986 release at Lake Nyos in Cameroon, in West Africa. Its waters were saturated with carbon dioxide and a major disturbance, most probably a landslide, caused a huge cloud of carbon dioxide to bubble up from its depths and pour down the valleys that lead from the crater. Carbon dioxide is denser than air, so the CO_2 cloud, traveling at 50 mph, hugged the ground and smothered everything in its path. Some 1,700 people were suffocated. In this event the CO_2 content was relatively small compared to the unthinkable volumes of gas that could be released from Lake Kivu.

Of all the diverse and intriguing solutions proposed to counter global warming, the most dangerous, because it is the most seductive, is this concept of carbon capture and sequestration (CCS). Most other plans provide a kernel of relief, in that they decrease the emission of warming gases into the atmosphere. Only CCS increases the amount of carbon that has to be dealt with. Once commenced, this program produces quantities that, if sequestered, would use up all of the Earth's naturally occurring underground spaces.

Epilogue

The objection to most processes that put CO_2 into the atmosphere or try to take marginal amounts of CO_2 out of the atmosphere or concentrate on putting less CO_2 into the atmosphere is that all accept the fact that less CO_2 is better than more CO_2.

But, to torture a tenet of architecture, less *is* more when it comes to the amount of CO_2 remaining in the air. Less CO_2 means only that the moment when the Earth becomes inhospitable is pushed forward a bit.

That is why it is vitally important to identify every last ounce of CO_2 that might be directed toward the atmosphere. To repeat, one-quarter of *any* present quantity of emitted CO_2 will still be in the atmosphere affecting the climate in 10,000 years. Unlike nuclear waste, where the quantity produced is insignificant compared to how long it remains dangerous, the problem of CO_2 is informed by the relationship of quantity to time, where the quantities of CO_2 are far from insignificant. Unlike highly radioactive nuclear waste, in which the tonnage is in mere thousands, there is simply no place to put billions of tons of CO_2 even if we were able to capture and recall the gas.

There are even good reasons to avoid not only sequestration projects but also conservation projects, since the Circe who whispers to us that we are saving energy and therefore putting less CO_2 into the atmosphere is a seductive, dangerous half-truth. The central and arguably immutable fact is that "less" does not begin to solve our two-hundred-year gluttony of carbon. At best, "less" simply makes us feel better. When the public feels better, it makes it possible for those in pursuit of personal profit, as our capitalist system requires,[49] to continue to emit CO_2 longer than they might have.

No program that seeks to marginally reduce present giga-sized outputs of greenhouse gases is going to make much difference in the quality of life of the immediate and intermediate generations to come. Indeed, even if we were able to turn a valve and cease producing any CO_2, it would have little effect in the near term. Climatologists tell us that damage has already been done by the CO_2 presently in the atmosphere.

So we are offered a very bad bargain. Continue to burn fossil fuels until they are exhausted and suffer sooner the degradation of our standard of living; or stop or severely mitigate the production and suffer just a bit later.

For those of us comfortable in the present condition, the decision could likely be short-term acceptance that the distant future will suffer but we and our immediate progeny will still be all right. But over the long term, if history is any indicator, our decision will, one hopes, be a moral one that will sacrifice present well-being in order to benefit generations in the future. We, our children and our grandchildren, should fight tooth and nail to put off the inevitable in the hope that, in some lucky future, we just might become smart enough technologically to alter future climate events that now appear to be our descendants' unhappy fate.

CHAPTER FOUR

Nuclear Waste and Friable Grids

The Fallacies of Nuclear Waste

SHOULD you find yourself in a discussion of the good and the bad of nuclear energy, the show stopper is reference to the figure 10,000. That number is what the antinuclear community usually quotes for the length of time that radioactive fuel remains dangerous. It is at the low end of the scale of doomsday, as some conjure with ten or twenty times that amount, and a few, in desperation to terrify, speak of a million years during which nuclear waste must be kept safe and sequestered.

The entire written history of the human race is barely 8,000 years, and a message from that distant past to us is rarely written in any language that we can understand today. So, looking forward, what is the likelihood that some distant descendant 9,000 or more years from now could comprehend the instructions that the stuff tucked away under Yucca Mountain must not be tampered with for yet another thousand years?

And who would guarantee that there would be anyone in authority able to read the message, let alone enforce an embargo? And what sort of containers must be developed to last 10,000 years, when the best we can do with stainless steel would be for some hundreds of years before leakage began? The truth is that those numbers do not represent the way the world works. A

Roman engineer in 100 AD did not plan for eventualities in the year 2000 AD. He more likely had the year 120 AD in mind. Nor, to come closer to home, did the industrialists of the nineteenth century, as they converted fossil fuels to global warming greenhouse gases, consider the CO_2 problems of even the twenty-first century, let alone the thousands of years beyond that, when the CO_2 that the nineteenth century manufactured will continue to affect climate.

Each generation inherits what is good and what is bad from the generation precedent and somehow makes the necessary adjustments for survival and growth. No one looks thousands of years ahead. No one *can* look thousands of years ahead.

In the case of nuclear waste, it is not two hundred generations into the future that we need to be concerned about but two generations hence—our grandchildren. Certainly we can arrange to protect a stash of radioactivity for a mere hundred years and leave the problem of the next hundred years to our grandchildren's grandchildren. And so on.

In any event, the problem of dealing with nuclear waste has already been solved. The truth is that there has not been very much nuclear waste generated. In fact, all nuclear waste represents less than 0.01 percent of all industrial toxic wastes. Since the beginning of the nuclear era, seventy years ago, a total of only 56,000 tons of nuclear waste has been generated by the U.S. nuclear industry. To put 56,000 tons produced in seventy *years* in perspective, consider that it is less than the 70,000 tons of waste generated each *week* by New York City. To emphasize the comparison of nuclear to coal, consider that "a 1,000 MWe nuclear power station...produces approximately 30 tonnes of high level solid packed waste per year if the spent fuel is not reprocessed. In comparison, a 1,000 MW(e) coal plant produces 300,000 tonnes of ash per year."[50]

Finally, in the mid term future, there are reprocessing technologies that recapture or reuse spent fuel and result in manageable low-radioactive residues. The long-lived dangerous radioactive material that extracted is by these processes is a tiny fraction of the initial weight of the radioactive waste. Some of the small nuclear reactors that the Nuclear Regulatory Commission is slowly beginning to consider either use nuclear waste produced from breeder reactors or reuse their own radioactive waste products as fuel. The bottom line is that if we reprocessed all of the nuclear waste that has piled up since the inception of the nuclear era, we would end up with less than 500 metric tons of the highly dangerous stuff. Certainly we can figure out how to sequester that amount of anything.

Long term, there is the suggestion that radioactive material can be transmuted into a stable form with no radioactivity. A firm in Belgium, SCK-CEN, which already produces 25 percent of the world's medical isotopes, has been given the go-ahead by the Belgian government to do serious work in this field. The plan is to start construction of a transmutation plant in 2015, with an initial operational phase starting in 2023. The budget has been set at almost $2 billion for this very preliminary work. It is hard to argue that transmutation, which presently is at least theoretically possible, is not in our future.

The same process is being investigated at Idaho State University, in cooperation with the Idaho National

Laboratory, where scientists are studying the possibility of entirely eliminating the "heavy" elements in waste, such as plutonium and americium. The project, begun in February of 2010, is being funded by the U.S. Department of Energy. Results leading to the elimination of all nuclear waste could emerge in the immediate few years.

A similar project is being developed at the University of Texas in Austin. Scientists there are working on a hybrid fusion-fission reactor, in which waste is held in place around the reactor core and destroyed by firing streams of neutrons at it. Daunting technical hurdles need to be overcome in order to fully optimize the device, but researchers predict that, once perfected, it could reduce nuclear waste by 99 percent.[51]

The promise of transmutation is the ultimate answer to the bugaboo of the millennial half-life of nuclear waste. If transmutation is possible, we would need only a onetime crash program to eliminate all existing nuclear waste that has accumulated since the beginning of the atomic age. Better yet, and going forward in the immediate future, we need not manufacture *any* long-lived nuclear waste. There are proven nuclear generating technologies that have been around for fifty years, some of which are presently operational, that would produce a quarter, and even less, of the waste that we now turn out, and the waste produced by these technologies would be dangerous for only 300 years, not the threatening 10,000. The use of thorium, an element of which you can probably find traces in your own backyard, instead of much more scarce uranium, could lead to a world of abundant nuclear power with no accumulation of eons-long dangerous nuclear waste.

We have real problems to solve in connection with nuclear energy, most of them financial and political, not technological. In what sense does terrifying the public over a problem, one element of which has long since been solved and which in the near future very well may be no problem at all, bring us closer to a solution of our pressing energy needs?

The Friability of Grids

Electric power grids that span continents exist for two reasons. The first is to transmit power to consumers from production points and the second is to balance out the varying demands that arise from widespread areas with differing momentary energy needs.

Transmission of power over long distances arose as a result of the increasing efficiency of scale of prenuclear generating technologies. The earliest power plants were located as close to users as possible to minimize power loss in delivery. Thomas Edison's first electrical generating plant, powered by steam, was located in lower Manhattan, where delivery was direct to nearby buildings. (Edison was among the first to recognize that the excess heat generated by his boilers had value. He piped hot water for space heating to buildings nearby.)

As coal plants became larger, they were necessarily sited away from the centers of population, where they had become air polluters and unsightly nuisances. The move away from consumers generated energy loss in transmission, but this was compensated for by the thermal efficiency of coal. As long as coal remains cheap, and that will be for a very long time, inefficiencies throughout the energy industries can be accepted. It is the low cost of coal that covers the 10 percent loss of energy in long-distance grid transmission that we now endure. Should the cost of fuel increase, the distances over which energy can be economically transmitted will shorten, and at some point transmission will become problematical. There is a price for fuel at which the energy industry might have to revert to siting similar to the intensely local Edison model.

The present grid was built in a time of innocence, before landowners discovered they had valuable rights, and long before the lawyers discovered that siting problems for anything associated with energy were a major source of income. The rights to build the present grid were cheaply bought. The rights to now extend, rebuild, or build anew a smart grid will not be so easily acquired. In preparing for the prospect of thousands of lawsuits, claims, and counterclaims, the Federal Power Act has promulgated strict rules limiting the traditional rights of landownership by individuals.[52] A further complication to building a new grid was a decision by the Supreme Court not to consider a lower court's opinion on transmission siting in which the lower court held that the states, not the federal government, had jurisdiction. This has produced a gridlock of monumental complexity. Dozens of parties now feel they have valid claims supported by the highest and the lowest courts in the land.

Thus the two sacred cows of American polity, states' rights and the protection of owners from governmental takeover of their land, are in play. These concerns are heightened by the fact that the present grid simply grew helter skelter, crossing all manner of governmental borders, jurisdictions, and, most important, private interests.[53]

One concern about providing renewable energy to population centers is the disjointed nature of the U.S. electric grid. In numerous instances, there is virtually no way to transmit power from the areas in which it can be generated to the places where it is needed and massive upgrades to this system are needed. However, as long as parochial decisions can disrupt the development of efficient transmission, the future of a comprehensive energy program in this country remains in serious doubt.[54]

The regulatory debacle adds interest to the question, to be dealt with in a later chapter, of whether we need a national grid at all or whether there is not some distribution model that allows the secure delivery of energy at a reasonable cost and with less economic and political disruption.

At present research and resources are being invested in developing the concept of a smart grid.[55] Smart gridists have gathered a host of desired capabilities under their banner and have suggested that many problems relating to the generation and distribution of energy can benefit through such a solution. But in addition to the enormous engineering and technological challenges it presents, there remains the matter of siting, as discussed above, and the problem of cost. Present estimates of the cost of a smart grid are at the $200 billion level, projected forward to 2030. Experience tells us, and prudence dictates, that a multiplier of five is not too much to allow for all of the possible problems we do not yet know about the smart grid concept and all of the as yet undone research and litigation that will be required.

The interesting part of a smart grid is that it would enable two-way communication between user and producer to balance out peaks and lows of demand, both within the grid and within the user's own house. This is certainly a good thing, but one wonders whether the game is worth the candle of cost and uncertainty. Not only do we not yet know what the cost of a nation-wide smart grid ultimately will be, but we do not yet know whether it will work at all and, if it does work, whether the savings will be substantial. If the savings are substantial, we also do not know whether a smart grid, or indeed any grid, can be kept secure.

Apart from the conceptual challenges of a new grid, the fact is that we need to retrofit our existing grid to add needed gigawatts of nuclear and other green power that will be required by ever-growing demand. One underreported segment of the energy picture is the draw that electric cars will make on the grid as the number of such vehicles increases from a thousand to a million on the road.[56] When we approach millions of electric vehicles, there is no grid that could reasonably accommodate them.[57] By that time either we will have had to eliminate altogether the electric grid as we know it or we will need to augment the national grid with intensely local sources of energy generation.

Storing electricity on a small scale is relatively simple. Batteries; pumping up water to higher levels; compressing air and spinning multiton flywheels—all are well within our present capabilities. But we must keep in mind the terawatt quantities involved in conti-nental and, ultimately, international storage of electric-ity. We are not yet dealing with this mammoth problem of the amount of stored electricity we will have to accommodate or with the transitory quicksilver quality of the energy to be stored. However, there is one new and untested proposal for both the delivery and the storage of large amounts of energy. We will deal with it later in this chapter.

As powerful and as technologically adept as we are, there are forces beyond our control that can affect any grid, no matter how high its IQ. The short-term effects of global warming are still not fully understood; in the near future we will begin to demand more giga-watts of energy to counter these changes. Climatologists suggest that weather will demonstrate peaks and lows of forces and temperatures uncharacteristic of the recent past. Outages from storms and extreme temper-atures have always been factored in by consumers, but at the level of a minor inconvenience. But when our lights grow increasingly dim or go out for long periods as a result of disturbances a thousand miles away, the benefits of an ever more sensitive grid dissipate and the search for standby power will intensify. In Florida there are already millions of small, highly inefficient diesel-powered generators in households awaiting the next hurricane. In its own way the Florida model works for the individual consumer while it is a disaster from the standpoint of ecological probity. It is a model that, in essence, eliminates a householder's need for a grid in the most needful moments. As we shall argue, it is a model for which we have at hand the tools and the tech-nology that will eliminate the need for a national grid by bringing generating capability directly to the consumer without further increasing the greenhouse effects of CO_2.

Weather aside, there is another eventuality about which we know very little and for which we are nakedly unprepared. In 1859, before there was much electrical equipment, let alone electronics, a flare on the sun played havoc with the infant electricity industry. The flare is known as the Carrington event, after Richard Carrington, an astronomer who was projecting an image of the sun onto a white screen when he witnessed the supersolar flare that set off the event. A high-intensity burst of electromagnetic energy followed the sun flare. It shot through telegraph lines, disrupting communications, shocking technicians and setting their desks on fire. The northern lights, energized by the flare, were visible as far south as Cuba and Hawaii. But in a non-electronic age, where all of the world's knowledge and all of its records were preserved on paper, not much lasting damage was done.

Today the vast and sensitive grids that we have constructed,[58] and the even more sensitive lines we are planning to construct, would instantly transmit an electronic holocaust that would wipe out everything from the warehouse-sized computers in our military to the cell phones we carry in our pockets. If the solar event was as powerful as the one in 1859, it would also ignite the pocket in which the phone was being carried. But most of all it would wipe out the grid itself. Such an event—whether naturally occurring or a man-caused disaster—could result in trillions of dollars in damage and claim more lives than were lost in World War II.[59]

According to cosmologists, another strike is highly likely. The resulting chaos could thrust the entire industrial world back some centuries into a pre-industrial model. Superflares are predicted to occur approximately every hundred years. We are already fifty years past this best guess.

Any wired grid, supermalt or plain vanilla, cannot be defended against periods of extreme solar flares.

Some idea of the damage of which a solar flare is capable was the previously mentioned Carrington event in 1921. Based on the experience of that event NASA made the following prediction:

> A contemporary repetition of the Carrington Event would cause...extensive social and economic disruptions. Power outages would be accompanied by radio blackouts and satellite malfunctions; telecommunications, GPS navigation, banking and finance, and transportation would all be affected. Some problems would correct themselves with the fading of the storm: radio and GPS transmissions could come back online fairly quickly. Other problems would be lasting: a burnt-out multi-ton transformer, for instance, can take weeks or months to repair. The total economic impact in the first year alone could reach $2 trillion, some 20 times greater than the costs of a Hurricane Katrina or, to use a timelier example, a few TARPs.[60]

Actually, we need not wait for a solar event to destroy the grid. The destruction could be accomplished by an electromagnetic impulse created by a force similar to the detonation of the primitive atom bomb named Mark 18, which was tested in 1952.

Today, if just one of these 500 kiloton bombs were detonated 300 miles above the central United States, the economy of the country would be essentially destroyed instantaneously. Very little of the country's electrical or electronic infrastructure would still be functional. It would likely be months or perhaps even years before the electric grid could be repaired because of the destruction of large numbers of transformers in the power grid, which, incidentally, are no longer made in the United States. Several countries today have the ability to produce a weapon similar to this 1952 bomb and send it to the necessary altitude. (England tested a single-stage weapon with a yield of 720 kilotons, called Orange Herald, on May 31, 1957.) The number of countries with this ability will undoubtedly increase in the coming years.

"The instantaneous destruction of the power grid would occur primarily because of the widespread use of solid-state supervisory control and data acquisition devices in the power grid. These would be destroyed by the E1 pulse, but could probably be replaced within a few weeks. The greater problem would be in re-starting the power grid. [No procedures have ever been developed for a "black start" of the entire power grid. Starting a large power generating station actually requires electricity.] The greatest problem would be the loss of many critical large power transformers due to geomagnetically induced currents, for which no replacements could be obtained for at least a few years. The loss of many of these power transformers would greatly complicate the re-start of the parts of the grid that could be more quickly repaired."[61]

And this from the National Academy of Sciences:

Power grids may be more vulnerable than ever.
The problem is interconnectedness. In recent years, utilities have joined grids together to allow long-distance transmission of low-cost power to areas of sudden demand. On a hot summer day in California, for instance, people in Los Angeles might be running their air conditioners on power routed from Oregon. It makes economic sense—but not necessarily geomagnetic sense. Interconnectedness makes the system susceptible to wide-ranging "cascade failures."

The only defense proposed by NASA has been a total temporary shutdown of the grid during predicted periods of solar activity. In this event, the cure would be as bad as the anticipated disease, because all electronic communication and control would cease, and for the hours or days that the shutdown lasted we would, for example, be scrabbling about to find the water to flush our toilets...and worse, much worse.

The odds against a terrorist attack on the grid, either by cyber means or by good, old-fashioned TNT, are not measured in centuries, as are solar flares. These attacks are measured in the headlines of yesterday and tomorrow. We know that cyber attacks are already being mounted. In June 2009 the *Wall Street Journal* reported attacks in which the intruders left behind software programs that could damage the grid.

"Espionage appeared pervasive across the U.S. and doesn't target a particular company or region," said a former Department of Homeland Security official. "There are intrusions, and they are growing," the former official said, referring to electrical systems. "There were a lot last year."

And a senior administration official confirmed that the Pentagon had spent $100 million in the last six months (2009) in repairing cyber damage to its network. The industry estimates that between 2010 and 2015 we will be spending billions in security dollars every year to keep our developing smart grids safe from cyber attacks. If past experience is any guide, the problem of security is essentially not solvable, so long as the grid is exposed and controlled electronically, and since our enemies are as smart as we are and are ever seeking more electronic chinks.[62]

Many of the present cyber attacks are by nations that seem to be preparing for the next war, when they would wish to fry the systems that make our economy work. A large-scale attack on the grid would essentially cost nothing in blood or treasure to the attacker. Seeking such a cheap advantage would, in the midst of a war, be impossible for an enemy nation to resist.[63]

Although we are not presently in a conventional war, we are faced with a much more immediate enemy, terrorists who need no large excuse to attack. One scenario is that a small group of terrorists could easily replicate the kind of cyber attack on our grid that any large nation might mount. Attacks could occur anywhere within the 160,000 miles of high-voltage transmission lines. (Energy experts are predicting that an additional 12,900 miles of indefensible grid will be needed over the next five years.[64])

While a national enemy would have to wait for the inception of a war to take out the Internet along with our grid, terrorists have no such burden placed on their plans. Any ragtag group of smart technicians with about a hundred thousand dollars' worth of computers could take down the grid and the Internet almost at will. Why they have not is a puzzle, but there is no mystery about whether they will take a shot. Why would they not?

To summarize, the smart grid is no smarter than our present dumb grid when the losses in energy transmission are considered. Any wired grid, smart or otherwise, will dump up to 10 percent of its initial power along the way. This is no small matter. Consider that if all of the energy wasted in transporting electricity over a grid were saved, it would be the same as if all the automobiles and trucks *in the world* had their greenhouse gases permanently removed.[65] Add to that the enormous costs of maintaining and expanding the grid and the unthinkable aforementioned problem of defending against natural or man-made attacks. The blindingly logical conclusion must be that distribution of energy over great distances from centralized large generating units is untenable in our electronics-dependent economy.

When all of the problems inherent in a national wired grid are taken into account, and when our lack of effective defenses is factored in, it is clear that the distribution of energy is not served well by an electrical, electronically controlled, wired grid. An entirely new way of thinking about the secure distribution and storage of energy must be considered.

The Special Relationship Between the Grid and Water Cooled Nuclear Power Plants

Since electricity cannot be easily stored, the primary function of a national grid is to immediately distribute the electrical energy to users. This is true for both fossil fuel plants and nuclear plants. However, there is a special relationship between the Grid and nuclear generation. The grid is the first line of defense against a nuclear accident which could lead to a meltdown since the grid is the source of electricity to power the pumps and valves that supply the coolant which would be required for a reactor approaching criticality. For whatever sequence of events that would lead to an emergency, the need for instant and substantial electrical power is evident.

In one case the grid itself could be the initiating cause of a nuclear emergency. With the Grid down, a nuclear plant having nowhere to dump its energy would be required to cease operation. In a shut down a nuclear plant needs power for extensive electrical driven water cooling since the core remains hot. To prevent meltdown it must be kept under water. In most plants in use today, stand-by power depends on diesel generation which becomes the last line of defense against a nuclear accident. Should the diesel back up fail, for whatever reason, a core meltdown would be almost inevitable.

This is a scenario of disaster waiting to happen. Any number of events, previously described in this book, could cause the grid to go down and require shut down of a nuclear plant. There does not have to be an accident in the plant itself to precipitate an emergency. The absence of the grid would be the initiating event.

An Alternative Grid?

In casting about for a solution to the indefensibility of a wired grid in an electronic world, a most unlikely player appeared.

Dave McConnell, an experienced engineer in the oil industry, retired in 1997, planning to put his savings into new oil drilling sites. Discouraged by the variability of oil prices, he began to think away from production of energy toward the larger problems of storage, security, and distribution of energy. He has proposed an entirely novel approach to an energy system, and he claims to have done so within acceptable financial parameters.

Recognizing the problems inherent in distributing electricity over wires, McConnell began to research the possibility of distributing energy by means of compressed nitrogen utilizing the vast network[66] of buried gas pipelines that spiderweb the United States.

In its simplest form, he proposes to pressurize pipelines with nitrogen at 1,440 psi and convert this pressure to the generation of electricity at sites close to consumers. A pressurized pipeline—think of a balloon—instantly delivers pressure that is potentially translatable to energy equally to all points along the line. McConnell's system, which he calls Lancaster Wind Systems, or LWS, accomplishes this and has other surprising and unintended advantages.

1. Instead of 10 percent loss of energy due to wire resistance, LWS purports to deliver energy over thousands of miles at a much lesser loss due only to pipeline friction.

2. LWS is a system that both delivers energy via pressurized nitrogen in pipelines and stores huge quantities for future use in the same container in which it is delivered.

3. A buried pressurized pipeline is not susceptible to solar flares. Its stored energy backs up elements of the system, such as turbines, generators, and control rooms, that would be affected.

4. A buried pipeline is resistant to the interconnectivity problems that could magnify a cyber or solar attack.

McConnell proposed pressurizing the system with the wind turbines now in use and eventually phasing them out and replacing the turbines with much cheaper, specially designed hydraulic windmills. However, since the system is not dependent on where or what kind of energy is injected, there may be a clear advantage in using small nuclear reactors to energize the system. These could be widely sited; think again of a balloon. It has been estimated that one 3 MWe nuclear reactor could keep a pipeline 750 miles long fully pressurized when in use as a storage facility. Small reactors, such as the mPower 100 MWe reactor from Babcock & Wilcox or the 25 MWe reactor from Hyperion, could be conveniently sited along the pipeline to supply 1,000 MWe of energy at any point in the line.

When the cost of building and siting ten small reactors, the smallest now hovering at about $100 million each, is compared with the cost of building and siting some thousands of windmills or the cost of building a 1,000 MWe maxireactor, there is no contest. Not only is the original construction cost in favor of nuclear, but the complexity of simply maintaining thousands of widely dispersed windmills and the decades required to build a maxireactor put a heavy thumb on the scales.

But the major economic argument in favor of taking a long and serious look at the pipeline proposal is that it already exists throughout the nation. If it proves technologically feasible, it would negate the need for the untold billions required to build and rebuild our electric grids.

A grid based on pressurized nitrogen pipelines is simply a proposal. However, it is a genuinely new idea, accomplishable by tested nineteenth-century technology. Because it delivers energy rather than electricity through its buried pipelines, it is resistant to most of the fragilities of and threats to an electric grid.

As such, it deserves serious consideration.

CHAPTER FIVE

Nuclear Hysteria

HARRY DAGHLIAN became the first fatality of the atomic era when a tungsten carbide ball fell from his hand onto a plutonium core on which he was working. On August 21, 1945, Daghlian was accidentally irradiated while performing a critical mass experiment on our first atomic bomb. Reflexively, he pushed the highly radioactive ball away from the core to stop the chain reaction he had accidentally initiated. In doing so, he received a radiation dose estimated at 510 rem (for Rontgen Equivalent in Man). He died twenty-eight days later.

Daghlian was the first of a total of only seven U.S. immediate fatalities recorded in the seventy years of the nuclear era. It is a stellar safety record unmatched by any other industry, nuclear or nonnuclear, in the world.

In the United States we have 30,000 deaths per year attributed solely to coal pollution. The hard truth is that anyone working in the nuclear field is 300,000 times more likely to die from the effects of coal pollution than from a nuclear explosion or radiation. The comparison becomes more absurd and impressive if we include all industrial deaths, not only those from coal pollution, and even more so if we include all of us who are not in the nuclear industry.

Since it is so convincingly, historically demonstrated that the chance that any American will die of a nuclear incident is so infinitesimally small as to be discounted entirely, it may well be asked: From whence does the virulent opposition to nuclear energy arise?

TRACED BACK to its beginning, opposition to all matters nuclear is rooted in the horrors inherent in the possibility of a nuclear holocaust. During the cold war and especially during the Cuban missile crisis,[67] the most ridiculous threat that the United States and the Soviet Union could come up with against each other was the concept of Mutual Assured Destruction, MAD, as it so aptly abbreviates. MAD, with tens of thousands of hydrogen bombs targeted at the populations of both countries, depended upon the shaky premise that our nuclear-armed enemy and ourselves were under the control of leaders who were not mad. A pretty crazy way to run a world.

Add to the unthinkable death toll in the warring nations the probability of a nuclear winter that would follow from the dust thrown into the atmosphere by the MAD exchange. During this long time that the sun does not shine, most of the world's population, deprived of food, would simply expire—if not worse, descend to unthinkable barbarism.

The pictures and reports emerging finally from Hiroshima and Nagasaki in the mid-1950s energized the antinuclear movement to a level of incandescence that shed more heat than light. For the next fifty years the memory of the possibilities of a nuclear holocaust remained fresh in a generation weary of all war, let alone a nuclear war. During this period, the majority of Americans violently opposed the construction and the continued operation of the hundred or so nuclear power plants in the United States. Siting of any nuclear installation became an overriding problem, as organizations such as Greenpeace and the Union of Concerned Scientists kept the horrors of radioactivity alive.

While the glow of victory initially clouded our memory of World War II, any such feeling of triumph was pierced by the pen of John Hersey. No single piece of literature has ever had so long a reach as his "Hiroshima," the article to which the *New Yorker* magazine dedicated an entire issue in 1946. Hersey described the horrifying damage to human beings caused by the bombing. Seen from the distance of thousands of feet, the mushroom cloud was a glorious act of destruction of a despised enemy. Seen from close up, in Hersey's graphic descriptions of the effect on people of exposure to radioactivity, the cloud lost its generality and became, in the eyes of most of us, evil incarnate, as individuals emerged from the anonymity of 100,000 dead. Hersey's reports of people facing the unknown horrors of radiation poisoning had the power to alert the conscience of the world. Most of us would maintain that moral judgment for half a century.

Hersey set the tone of rejection of all matters nuclear, a rejection that was soon reinforced by another work of art, this time an apocalyptic account of a war between nations armed with nuclear weapons. *On the Beach*, a novel by Neville Shute, was released in 1957 in the midst of the cold war. The book and, two years later, the film dramatized the futility of any defense against radiation released by a nuclear exchange. The hopeless depiction, especially in the film, of the end of human history demonstrated the immense power of fiction when it is released onto prepared ground.

The artistic antinuclear drumbeat continued when, in 1964, satire nailed the truth in the film *Dr. Strangelove or: How I Learned to Stop Worrying and Love the Bomb*. Fiction though it was, the film was a recognizable image of how close we came to a nuclear Armageddon and how friable are our political structures. The image of the hat-waving Texas cowboy riding the end-of-the-world bomb down onto Moscow is ineradicable.

China Syndrome was released in 1979. Until this film appeared, all of the antinuclear artistic efforts had been focused on war and the bomb. For the first time in a popular medium the emphasis was on the eventualities of an accident in a civilian nuclear power plant. And, in an almost unthinkable coincidence, just twelve days after the release date America's first accident at a civilian nuclear power plant occurred. On March 28, 1979, half the fuel in one of the reactors at Three Mile Island melted down. That no one inside or outside the plant was exposed to lethal radiation; that the melted fuel did not proceed on downward to China, as suggested in the film, had little effect on the antinuclear hysteria that followed. The two events, so close in time, identical in image, powerfully reinforced the pandemic antipathy to anything nuclear. The stage was being set for growing and powerful resistance not only to nuclear war but to the use of nuclear sources for energy.

Dozens of public demonstrations against the nuclear industry were mounted during this period. On June 12, 1982, a million people turned out in New York City. It was, and remains, the largest political demonstration in American history.

At this moment any remaining hope for the development of safe nuclear power was crushed when a nuclear accident in the Soviet Union, further sensitized the world to the terrors of radioactivity.

On April 26, 1986, at 01:23 a.m. Soviet time, reactor number 4 at the Chernobyl plant, near Pripyat in the Ukrainian Soviet Socialist Republic, exploded. The disaster that resulted was considered to be the worst nuclear power plant accident in history. It is the only level 7 event on the International Nuclear Event Scale and resulted in a release of radioactivity following an uncontrolled nuclear chain reaction that destroyed the reactor. Most fatalities from the accident were caused by radiation poisoning. Further explosions and the resulting fire sent a plume of highly radioactive fallout into the atmosphere and over an extensive geographical area, including the nearby town of Pripyat. Four hundred times more fallout was released than had been by the atomic bombing of Hiroshima.

In spite of dire predictions of widespread deaths and illness from the disaster, no one outside of the Russian plant immediately died, was sickened by released radioactivity, or suffered trauma save only the thirty-two workers who were in the reactor containment at the time of the accident and the ten children who, as of this date, have died of thyroid cancer.

America and the world had been so terrorized by fiction that when events such as TMI and Chernobyl were quickly and safely contained, there was little impact on the unyielding public mind-set against nuclear power.

Until the developing crisis of global warming in the opening years of the twenty-first century focused the world's attention on how much more damaging the effects of fossil fuels were over that of nuclear fuels, not one new nuclear power project was either planned or built.

The powerful fictions of endless death and destruction combined with actual nuclear accidents had cost the world precious decades in the battle against global warming. For generations, the same forces will cost trillions of dollars to counter. In the meantime, we continue to dump carbon into the atmosphere, instead of developing crash programs like the Manhattan Project that could do much to counter the damage of too much carbon in the atmosphere.

CHAPTER SIX

Big Nukes, Big Problems

"If I had my druthers, I would say, Let's take out a clean sheet of paper, and let's design a new category of reactor that avoids all difficulties and is inherently safe."

—Dr. Alvin Weinberg, 1995

WEINBERG WAS THE nuclear scientist who midwifed the nuclear age in the 1940s, mothered it through the good years and the bad, and died in 2006, just as his "druthers" were beginning to look not only eminently sensible but a lot like the ideas of many other scientists and industrialists.

Today we are the inheritors of a nuclear technology that the same Dr. Weinberg attributed to a military "infected with gargantuism." The hunger of our armed forces for plutonium during the cold war for nuclear weapons concretized an energy technology that dealt in volumes of potential destruction so large as to scare the good sense out of most of the world. We are saddled with a technology that is too risky, too big, too costly to build, too slow in the building, too slow to dismantle, too uncertain, and totally resistant to change and future technological progress in any reasonable time frame.

But as Alvin's son, Richard, said, "Now that there is a technology...that works, there are huge barriers to anything new."

Today, regulatory approval for even a traditional LWR power plant, whose design goes back fifty years and of which there are a hundred in operation in the United States, could take five years or more to obtain. Once approved, it could take five years or more to build and $10 billion[68] to complete. The risks are not only financial.[69] The biggest risk is that during the decade required, and certainly during its half-century life, something better is likely to have come along, rendering the project obsolete or at least financially unviable. Already, proponents of small nuclear reactors are predicting that ten 100 MWe reactors would come in at a fraction of the cost, and a fraction of the risk, of an equivalent 1,000 megawatt generator.

Estimates of future demand for energy indicate that by the year 2030 the world would require the output of an additional three hundred 1,000 MWe reactors. On the basis of present U.S. energy use, which is about one third of total world use, the United States will need approximately one hundred big reactors in the next twenty years. The prospects of building and paying for these in two decades is dim. The Obama administration is hoping to have only seven to ten reactors built in the next ten years. If this rate holds, we may have built approximately thirty reactors when in 2030 we will need a hundred. Considering all of the proven difficulties of siting and financing and regulation, those desperately required gigawatt plants are not likely to appear.

At the end of the normal life of a 1,000 MWe reactor, provision needs to be made for the billions of dollars that will be required to deactivate and clean up the site. And should the reactor, any time in its life, suffer even a minor nuclear accident, the cost could climb to billions times ten. There are no financial models that can predict or even approximate risk rates that are projected so far into the future, with so many parameters and extreme exigencies possible. The result will be that the public will pick up much of the capital costs in building the megawatt units. More threatening will be governmental exemptions of liability by the owners of nuclear plants. This will impose enormous burdens on the taxpaying public as the insurer of last resort.

Add to all of these hurdles the scarcity of acceptable sites, especially the most rare land, waterfront land, desired by megawatt developers. Add to that the enormous burden on water supply,[70] and the fervent NIMBY opposition, and one begins to wonder whether the future supply of energy from the huge 1,000 MWe reactors will be able to keep up with the burgeoning demand for energy from existing populations, not to mention the 50 percent increase in population expected over the next few decades.[71]

To add to the problem, many of the world's nuclear reactors burn highly enriched uranium, which is increasingly problematical especially in an age of nonnational terrorism. The threat of proliferation of potential bomb-making material requires that present and future reactors all use safer technologies. Old reactors can be converted to low enriched uranium even though there is no guarantee that the same performance would be achieved. And merely to design and test for a shift to safer fuel would take five more years.

Large-scale reactors are producing radioactive waste at the rate of 2,000 tons a year. Not one pound of this waste generated over the past half century has yet found a permanent and acceptable home. Rather, it is being held in temporary pools at the plants that produce it until the acceptance of safe sites (such as Yucca Mountain in Nevada) by affected communities can be worked out. There is no indication to date that this is any more possible now than it was a decade and $38 billion ago, when the stalled Yucca Mountain project commenced.

The opposition to burying radioactive waste over long periods of time is a problem of public perception rather than one of technology, yet there does not seem to be any compromise on the issue. Opposition is adamant, and since the solution, at least in the United States, essentially depends on local politics there is little will for resolution.

The most dangerous condition imposed upon our economy by huge centralized generation of electricity is likely to be the electric grid that is necessary to gather in and distribute plant output, whether it be from a nuclear or a fossil fuel source. As discussed elsewhere, the grid is the natural outcome of the centralization of generating capacity. It is an outcome that is unsafe due to the dangers and uncertainties of outside events that are capable of bringing the entire network down. So long as we continue to centralize the generation of power, we will remain hostage to cataclysmic events over which we have little control. The solution to this and to most of the problems inherent in our present profile of generation is to decentralize and to scale down reactor sizes. Decentralization of energy generation would, among other advantages, diminish the need for continent-spanning, delicately interconnected grids.

When all of these problems and restrictions are considered, it becomes clear that the future of nuclear power comes down to a matter of scale. Even without the introduction of new and inherently safe nuclear technologies that produce much less waste, a simple reduction in the scale of reactors would ameliorate at least some public opposition as well as ease the pressure on scarce water and location resources.

But we continue to build the big ones and resist the possibility that small, modular nuclear energy sources might just be the answer to problems of scale within the industry as presently conceived. The bureaucratic rigor mortis of the NRC does not yet allow for any consideration of a replacement technology before the next five years, which would stretch into decades before final approval.

Prior to the realization of the imminence of global warming, we had plenty of time, centuries, in which a leisurely and bureaucratic decision-making process was acceptable. We no longer have the time for consideration or the time for building sufficient 1,000 MWe plants to replace the fossil fuel plants that are at the core of global warming.

These small-unit, modular, inherently safe reactors can be built on an assembly line at high speed and shipped by the thousands on semi trailer trucks. Such plants are on our technical horizon, at levels of risk and pressures on resources that are orders of magnitude less than those produced by the behemoths into which we seem to be locked.

The French Experience

The French nuclear energy program depends entirely on giant pressurized water reactors (PWRs). The primary reason it is functioning so well is that the French commenced thinking about energy problems in 1946, when they simply nationalized their electrical generating industry.

The European left was strong and dominant at the time, and there was a rush all over Europe to nationalize basic industries. Since France was deficient in energy resources compared to other nations, the move to organize energy production in the public interest was natural and emerged from the need for both self-defense and self-sufficiency. The French energy industry at the time was a dystrophic collection of thousands of private interests, most structured as local monopolies with no motivation toward maximizing the collective good. The view of the politically dominant left was that private capitalism, which had failed Europe in the depression of the 1930s, and which had become the financial handmaiden of the Nazi movement, was both invidious and unworkable for the energy requirements of France. As a result, there was little opposition to the creation of the entirely state-owned entity Electricité de France, or EDF.

In the 1960s, farsighted technocrats in the EDF built France's first nuclear power plant, a modest PWR producing less than 70 MW of electricity. In the 1970s, in early recognition of the importance of nuclear energy, the Mesmer Plan was introduced with the political and financial backing of the French government. In 1974 and 1975, three new plants—Tricastin, Gravelines, and Dampierre, with a combined capacity of nearly 11,000 MWe—were begun. By 1977 work had begun on another five stations, all using PWR, with a total capacity of 13,000 MWe. Within a few years Superphenix, a fast-breeder reactor at Creys-Malville, on the banks of the Rhône, was begun. This reactor, using enriched uranium and plutonium, has the astounding capability of generating 12,000 MWe.

The United States at this time, in a shortsighted response to public opposition, was halting nuclear construction. The U.S. nuclear footprint began to decline as the French involvement in nuclear energy expanded. America fell decades behind while the French built fifty-nine plants, which, by 1990, supplied 75 percent of the country's electricity.

The EDF gradually expanded to include all energy sources as well as controlling the distribution of electricity over the national grid. By the turn of the millennium France had an effective and integrated energy policy while the United States was flailing about in a sea of oil, coal, and natural gas it was the major contributor to global warming.

It must be remembered that in spite of the obvious salutary advantage that the French enjoy in carbon free energy, they remain nakedly exposed to cyber and sun flare attacks on their country wide electric grid. Furthermore, since they have essentially no power sources other than nuclear, and that potential is concentrated in huge megawatt reactors, they remain at a huge disadvantage in terms of national security.

When, around the year 2000, most climatologists began to identify the CO_2 emissions from the burning of fossil fuels as a threat to mankind, France was relatively simon-pure in carbon production.

The reasons the French energy program may not work in the United States are many, but chief among them, because of the imminence of global warming, is lack of time. We do not have the luxury of time that was available to the French.

Other than the time problem there are those who ask why, if the French could do it, we cannot. The answer is that there is not the slightest chance, no matter what the threat, that we could nationalize our oil, coal, natural gas, and nuclear industries and take over the energy grids that span the nation. It was the ownership and control of all aspects of energy production combined with the federal power to integrate and rationalize the system that made it work in France.

Both state ownership and imposition of economic policies by force are antithetical to the American sense of who we are. We are both better off and worse off for our instinctive rejection of federal ownership of a means of production.

We need to find another way.

CHAPTER SEVEN

The Answer

Multum in Parvo

"The economy of scale is offset by the economy of mass production, no need for periodic refueling and the unnecessary need for redundant safety systems."

—Dr. Robert Sadlowe, Oak Ridge National Laboratory

Mini and Many

WHEN MR. ROLLS and Mr. Royce joined up at the turn of the nineteenth century, Royce, a working engineer, had long been thinking about the fragility of early automobiles. Things kept coming apart (the terrible roads did not help), and any breakdown in a machine that was essentially hand-built meant that the owner faced a very long time without his car.

Royce observed that most problems arose not in the elements themselves; the problems occurred usually where elements were bolted together. He further observed that car builders used a few very large bolts, which, he decided, was exactly the wrong thing to do. His assemblies would be bolted together with many small bolts rather than a few large ones. As a direct result of this observation, the foundation for the unassailable superiority of the Rolls-Royce automobile was set. There were fewer catastrophic failures, and when one of his small bolts showed signs of failing, or indeed did fail, replacement did not affect the entire system.

The parable of Rolls-Royce exposes precisely what is wrong with the track that the nuclear industry has been following for over half a century. The huge gigawatt breeder reactors, born of the military's need for fissile plutonium during the cold war, became the standard for the industry and, more importantly and casting a longer shadow, the standard for governmental regulation. Regulators embraced the familiar and suspiciously denigrated anything new in the industry. As a result, Royce's inherently wise advice—"many little bolts are better than one big one"—was ignored.

As we shall see in the following discussion of the problems associated with the nuclear experience everything, every problem, every option, and every solution, revolves around scale. Big is dangerous, big is scary, and big is infinitely more complicated and thus infinitely more prone to unforeseen consequences. Gigawatt chunks of energy, taken periodically out of service for maintenance, require expensive and polluting backup capacity. In addition, big requires huge dollops of scarce money and huge dollops of even more scarce time. Such daunting commitment of time and money sets up huge barriers to inevitable technological improvements in nuclear energy science that are already clogging our industrial pipelines. Commitment to gigawatt reactors is like a snowball rolling down a hill gathering strength, gathering mass, increasingly resistant to diversion and ultimately becoming unstoppable.

Our present nuclear industry is such a snowball. The industry is building on a process that has lost much of its creative dexterity because of its sheer size. Each year, as another arthritic monster comes on line, the concretization of existing energy design becomes more entrenched.

Siting and Inherent Grid Problems

Let us consider first the predicted need for energy for the balance of this century. It has been estimated that if we would supply all of America's future energy demands for the next fifty years exclusively from nuclear power, we would need to build and site five hundred 1,000 MWe nuclear installations. That would average about ten per state, distributed according to population and industrial concentration. Texas and California would need eighteen each, and Rhode Island would have to make do with one.

The problem of finding five hundred sites acceptable to the public, with sufficient water and land, would be a heavy task, but let us assume that we can, and that all five hundred are pumping out the power needed, and all are necessarily linked in one great grid so that each supports all of the others in case of weather events, maintenance needs, or even small accidents that might take out a plant or two. There are, however, other insurmountable problems inherent in this scenario.

Grid Transmission Loss

Such a scheme would suffer an overall loss of approximately 10 percent[72] of the power generated due to wire resistance and heat loss from transformers, those ubiquitous electrical boxes that hang on telephone poles in every neighborhood. That may not sound like much, but it's enough juice to run fourteen cities the size of New York.

Grid interconnectivity and Power Outages

The very element that makes such a system work, its interconnectivity, arrives with a pair of Achilles' heels. The left heel would be regional power overloads, which might or might not be able to be isolated in time and which have been responsible for enormous blackouts affecting millions of users in the recent past. The last serious blackout had an estimated price tag of $56 billion. The right heel is the probability of a solar flare that would take out the entire grid, fry all attached electric and electronic components, and thrust the United States back deep into the eighteenth century. A serious solar flare would lead to the disruption, or even the absence, of absolute necessities such as food and water for millions of Americans for a very long period of time. Lack of communications and the inability of any central government to exercise control and direction would lead to the tribalization of governance and perhaps to the end of our democratic tradition. Rebuilding to anything approximating our present standard of living and our present form of economic organization would take decades, if not longer. Forcing the rate of reconstruction by fiat, and all of the authoritarian methodology that implies, might take a direction far from the nature of government and economics that existed prior to the flare.

Cyber Attacks

Whether it is a lone hacker seeking thrills or a foreign government seeking industrial or military information, the attacks on the national grid have already begun. The U.S. government has allocated millions to ward off or at least to identify such incursions, but the attacks continue. The situation with reference to the present grid is bad enough, but with the proposed smart grid, with its two-way street and its interconnectivity with millions of devices, the problem has gone viral. The complexity of the problem of protecting a smart grid raises the national security question of whether it can be protected at all.[73] But the risk attendant on not making the enormous but unproven effort required is best understood by a statement from Richard Clarke,[74] who has been involved in cyber security for over a decade.

> The U.S. military is no more capable of operating without the Internet than Amazon.com would be. Logistics, command and control, fleet positioning—everything down to targeting—all rely on software and other Internet-related technologies. The logic bombs in our grid, placed there in all likelihood by the Chinese military, and similar weapons the U.S. may have or may be about to place in other nations' networks, are as destabilizing as if secret agents had strapped explosives to transmission towers, transformers and generators.

What applies to the military applies even more so to the less protected usages of the grid by civilian industrial and communication networks.

Cost Advantage

These problems, inherent in very big nuclear power plants, lead to the consideration of an alternative system of transmission, such as using 100 MWe SMR reactors, widely distributed, which would require no national grid and would present fewer siting and water problems. Because there would be no national grid to waste power in transmission, we would only need 4,000 SMRs to match the effective and eventual level of usable output of five hundred 1,000 MWe plants.[75] The cost of 500 of these monsters at a low estimate of $10 billion each is $5 trillion, while 4,000 100 MWe plants, at a generous estimate of $500 million[76] each, would cost a manageable $2 trillion.

Global Warming and the Pressure of Time

Far beyond cost is the problem of time. Our present level of CO_2 in the atmosphere rises every day, and finding a way to call a halt to these emissions means much less expenditures and much less trouble in the future. A 1,000 MWe reactor from the day of proposal to the day that electricity flows can take ten to fifteen years. Regulatory permission alone can eat up a third of these years, and no one has yet been able to bring in a large reactor within the time predicted. Each reactor is considered a separate and unique installation from the point of view of the regulators, even though the basic and previously approved design may be the same for all. The problem of standardization is confounded not only by technology but by site, population, and resources such as water for cooling.

Compare this to the proposals for 125 MWe reactors such as the mPower units designed by Babcock & Wilcox. The first of these, still on the drawing board, can be delivered, according to B&W, within a five- to ten-year window dating from the day of application to the NRC to delivery from what is essentially a factory assembly line. The second and the fifth and the tenth reactor would take progressively less time, and by the hundredth the reactors could start rolling out like tanks in wartime. At any moment that an incremental improvement was found, the assembly line would allow instant adjustment.

An interesting comparison relating to construction and delivery of an mPower or similar reactor would be to that of delivery of a Boeing 747. Both are subject to rigid regulatory oversight, and both are complex assemblages of electric, electronic, and hydraulic systems. The 747 is assembled from 4.5 million parts, probably far more than in a B&W 125 MWe reactor.

According to the latest figures, a 747 costs about $250 million, comparable to mPower when mPower is being built in the same mass production mode as the 747. Production of a 747 takes seven months from concept to completion. Since the planes are built on an assembly line, Boeing claims that once the assembly line is started a 747 is completed from a pile of parts to exit from the hangar in only four days.[77]

If we accept these estimates, it seems likely that B&W alone would be able to build and deliver the 3,000 small reactors by assembly-line methods within five years instead of three hundred 1,000 MWe reactors that would take decades longer and billions more. Additionally, other small reactor builders would be in production during the same period. Hyperion, for example, estimates that individual manufacture and assembly would enable them to turn out one 25 MWe reactor a month. If Hyperion reactors were to be mass-produced, we might well be seeing an assembly line that produces a Hyperion a day.[78]

If we limit ourselves to 1,000 MWe reactors, there is no way to approximate seriously the length of time it would take to build all five hundred of the reactors that we need. The year 2100 could easily come and go long before their completion. By that time, considering only population increase, the demand for energy would certainly rise above historic levels. Indeed, it might well simulate the apocryphal problem that the owner of a gas-guzzling SUV might have at the gas pump when he is asked by the attendant filling his tank, "Please turn off your motor, you are gaining on me."

In this nuclear age we now parallel the same condition that the U.S. auto industry faced at the turn of the nineteenth century. If Detroit had continued to build cars one by one by hand, America would look unrecognizably different today.[79] If we continue to build reactors, one by one, in a time frame inconsistent with rising demand and rising world temperatures, America as well as the world will certainly have a warming future.

The trend in all industry in the modern age is toward the spectacular speed of which a mechanized assembly line is capable.[80] The explosive growth of the computer industry, with its technology doubling memory every two years was made possible by the application of robotics to complex assembly. If we do not start designing reactors, keeping in mind from the beginning that they be built by assembly-line methods, we will continue to pour ever more resources into energy production with ever diminishing returns.

Desalinization and World Water Shortage

In 1990 Florida, Georgia, and Alabama began fighting over scarce water rights from increasingly scarce sources. After twenty years of fighting over the water from Lake Lanier, the sole source of most of the potable water for Atlanta, the city lost a federal court decision and now faces the daunting task of finding six hundred million gallons of potable water a day that just might not exist.

This is far from an isolated case in the United States, as states and municipalities are loading increasing demands on limited supplies of water. The battle in the West has so far been contained within the courts.

The third world has long been rent in recent droughts, by the search for water. In subsistence economies, on marginal land water is not a convenience but a matter of life and death. As a result small wars have been fought, rivers diverted, and wells poisoned in what could be a warning of what is to come as industrialized nations begin to face failing water supplies.

Quite aside from the demand for potable water is the dependence of enormous swaths of industry and agriculture on oceans of water used for processing, enabling, and cleaning a thousand processes and products. It is interesting to note that fresh water used in both industry and agriculture is reduced to a nonrenewable resource as agriculture adds salt and industry adds a chemical brew unsuitable for consumption.

More than one billion people in the world already lack access to clean water, and things are getting worse. Over the next two decades, the average supply of water per person will drop by a third, condemning millions of people to waterborne diseases and an avoidable premature death.[81]

So the stage is set for water access wars between the first and the third worlds, between neighbors downstream of supply, between big industry and big agriculture, between nations, between population centers, and ultimately between you and the people who live next door for an already inadequate world water supply that is not being renewed. As populations inevitably increase, conflicts will intensify.[82]

It is only by virtue of the historical accident of the availability of nuclear energy that humankind now has the ability to remove the salt and other pollutants to supply all our water needs. The problem is that desalination is an intensely local process. Some localities have available sufficient water from renewable sources to take care of their own needs, but not enough to share with their neighbors, and it is here that the scale of nuclear energy production must be defined locally.

Large scale 1,000 MWe plants can be used to desalinate water as well as for generating electricity. However we cannot build them fast enough to address the problem, and, if built they would face the extremely expensive problem of distributing the water they produce. Better, much better, would be to use small desalinization plants sited locally.

Beyond desalination for human use is the need to green some of the increasing desertification of vast areas such as the Sahara. Placing twenty 100 MWe plants a hundred miles apart along the Saharan coast would green the coastal area from the Atlantic Ocean to the Red Sea, a task accomplished more cheaply and quickly than through the use of gigawatt plants.[83] This could proceed on multiple tracks wherever deserts are available to be reclaimed.

Leonard Orenstein, a researcher in the field of desert reclamation, speculates:

> If most of the Sahara and Australian outback were planted with fast-growing trees like eucalyptus, the forests could draw down about 8 billion tons of carbon a year—nearly as much as people emit from burning fossil fuels today. As the forests matured, they could continue taking up this much carbon for decades.[84]

The use of small, easily transported, easily sited, and walk away safe nuclear reactors dedicated to desalination is the only answer to the disproportionate distribution of water resources that have distorted human habitation patterns for millennia. Where there existed natural water, such as from rivers, great cities arose and civilizations flourished. Other localities lay barren through the ages. We now have the power, by means of SMRs profiled to local conditions, not only to attend to existing water shortages but also to smooth out disproportionate water distribution and create green habitation where historically it has never existed.

The endless wars that have been fought, first over solid bullion gold and then over oily black gold, can now engulf us in the desperate reach for liquid blue gold. We need never fight these wars again as we now have the nuclear power to fulfill the biblical ability to "strike any local rock and have water gush forth."

Radioactive Waste

The original designers of the light water reactors presently in operation gave little thought to the eventual problems of the nuclear waste they produced. As a result, we now live with a problem that could have been entirely or essentially avoided in the decades when the nuclear power industry designs were being developed. The thousands of tons of nuclear waste that exist are being added to each year by 8,000 more tons created by the 436 gigawatt-sized reactors operating worldwide. This is an unavoidable and inconvenient fact with which we will have to deal.

Additionally, we are looking at an enormous demand for energy in the next fifty years. Based on projections issued by the World Nuclear Association in 2007, we will need, at the lower end of the projection, a worldwide total of 1,300 1,000 MWe reactors to service that demand. At the higher end of the projection, we will need 4,500 such reactors. If we continue to build the same gigawatt reactors that are now under consideration by the Nuclear Regulatory Commission, it would mean adding between 24,000 and 80,000 tons of high-level waste each year to our stockpiles of nuclear waste. This is clearly an unsustainable future for the nuclear energy industry.

On the other hand, every serious new design proposed for small reactors included the problem of waste as a factor in design from conceptualization. Solutions run from small reactors that diminish the amount of waste produced, to reactors that guarantee safe storage over time of diminished amounts, to reactors that produce no waste at all. Of the multiplicity of designs for mini modular reactors that are in play, what follows is how a few of the new SMRs deal with the problem of nuclear waste.

LFTR (LIQUID FLUORIDE THORIUM REACTOR)

One proposed reactor, the liquid fluoride thorium reactor (LFTR), worked on originally by Alvin Weinberg and Edward Teller around 1950, produces about one one-hundredth the long-lived spent fuel of our presently operating 1,000 MWe light water plants. That means that, had we been operating LFTR plants in the 1950s, as Weinberg clearly preferred, we would be dealing with 500 tons of waste rather than the 50,000 tons we have accumulated.

MPOWER

The B&W mPower 125 MWe reactor requires refueling every five years over an entire lifetime (sixty years). All of the waste material that the reactor generates is stored on-site in the reactor pool for the lifetime of the reactor.

PBMR (PEBBLE BED MODULAR REACTOR)

In the pebble bed modular reactor, the silicon carbide coating on the fuel spheres, the "pebbles," will keep the radioactive decay particles isolated for approximately a million years. All that is necessary is that the pebbles, which are stored dry, be kept under 1700 degrees Celsius. Prior to very long-term storage, the design of the PBMR allows for expended fuel to be temporarily stored for eighty years on-site. After eighty years, or during any of these years, the spent coated fuel can be disposed of in a deep underground repository. A coated pebble will maintain its integrity for far longer than even the decay time of 250,000 years of plutonium.

HYPERION

After seven to ten years of unattended operation, the Hyperion Power Module, capable of supplying all the electricity required for 20,000 homes, produces radioactive waste about the size of a basketball.

TWR (TRAVELING WAVE REACTOR)

TWRs proposed by a new firm, TerraPower, use a small amount of enriched uranium (10 percent enrichment) at the beginning of the process. After that, the TWR runs on nuclear wastes that have been accumulating for half a century. Using waste uranium from other reactors would reduce the total amount of nuclear waste in the world and considerably extends the available supply of the world's uranium.

The radioactive waste problem is being addressed in various ways by all of the developers of small nuclear reactors. The radioactive waste in these new technologies is a fraction to none of that required by reactors presently in operation. It can be demonstrated that almost any of the new designs could, in a few years, have the whole problem of waste well in hand. A public mythology has grown up around the idea that sequestration of enormous amounts of waste is dangerous and impossible. With the new mini reactor designs, massive sequestration will not be necessary. Changing public perception concerning waste will not be easy, but the facts of the new nuclear technologies cannot be denied.

The course that is the most dangerous to our future and most expensive in current dollars is the thinking that dominates the nuclear industry as presently conceived: namely, continuing to build the factories of waste that we have been proliferating for over half a century.

The Problem of Regulation

The U.S. Nuclear Regulatory Commission is thirty-five years old. Over those thirty-five years, the NRC has not certified any new reactor types. Today, only four designs can be referenced for applications to build and operate a nuclear power plant.[85]

During all of these years, the mood in the country was essentially antinuclear. As a result, the NRC concentrated on the problem of safety of existing plants rather than looking to innovation. The thinking was that since no new nuclear plants were going to be built (in fact, some were closed down by popular clamor during these years), why have staff capable of looking at new approaches to nuclear power? As a result and by its own admission the NRC simply has not had on board for decades the level of expertise required to quickly and safely approve innovative, problem-solving designs. It was, and remains, a closed and restrictive approach to energy supply in an industrial world that has proven that open and receptive regulatory mechanisms enhance and encourage innovation.

By 2000, when the CO_2 problem pointed up the nonsustainability of burning coal, oil, and natural gas, the NRC still had no staff capable of even evaluating new reactor designs. Considering the antinuclear mood at the time, this is neither surprising nor does it bring dishonor to the NRC. What does bring dishonor and even dismay is that now, a decade later, when it has become clear that the mood of the country is changing toward civilian nuclear electrical generation, there is still no adequate cadre of NRC engineers capable of moving along the licensing of the new small nuclear reactor models emerging from our laboratories and our universities. In fact, at a panel discussion of small reactor vendors, Michael Mayfield, head of the Office of Advanced Reactors at the NRC, announced that the NRC is "unfamiliar with most small reactor technology and has no proven review process to certify one."

Having refused to accept applications for licensing, still the NRC has accepted seven SMR models[86] for "precertification review," but has nothing to say about when applications will be accepted or anything at all about the baker's dozen of other innovative models that are being developed and were not accepted for precertification. The best guess, gleaned from various statements leaking out of the NRC, is that it will be five years or more before applications for SMRs can even be accepted and another five years at least before licenses can be granted. In fact, the TVA has recently announced that it wants to build an mPower reactor for its own use but, due primarily to the regultory process, TVA figures that the reactor cannot become critical in the next twenty years. This kind of regulatory delay serves to stifle entrepreneurial activity and, much worse, it dissuades the accretion of capital required for continuing research and development by requiring that the NRC be paid by applicants for the costs of licensing.

Smaller plants like those presently in development are severely handicapped by the fact that NRC licensing cost tables are computed on a per-plant basis, without any discounting for reduced size or complexity. Innovative ideas are also handicapped by the fact that Nuclear Regulatory Commission expertise is clearly limited to the fifty-year-old technology of existing light water reactors. The current approach for any developer of a new reactor is to pay NRC employees approximately $200 per hour each to learn something they already know.

The emergence of small nuclear reactor designs changes the nuclear game since a 100 MWe *inherently safe* reactor eliminates both the perceived and real dangers of a 1,000 MWe reactor which is merely engineered safe. It would seem, then, that the regulatory process that deals with the safety of the new generations of SMRs should take far less time and money.

Furthermore, the regulatory process will be much less onerous, as most of the SMR vendors are proposing a modular profile. Thus, after the initial model receives approval, the next one or the next one hundred SMRs of a particular vendor would require much less additional investigation and regulatory oversight. Or so that would seem in a rational world but, as regulations stand, now each exact replication of a modular reactor would be required to go through the entire regulatory process.

Proliferation

Part of the nuclear waste being produced worldwide by existing large-scale breeder reactors consists of weapons-grade plutonium and neptunium. In 2009 alone over 120 tons of this most worrisome of all nuclear wastes was added to the stock of weapons-grade material that has been accumulating for decades. As we add the thousands more gigawatt reactors that will be required in the balance of this century for energy needs, this stockpile of bomb material, which remains dangerous for hundreds of thousands of years, will grow apace. As more and more countries quite properly demand the benefits of nuclear energy, the storage of material useful in the making of weapons will be spread uncontrollably around the world in the hands of ever less politically stable governments.

This is clearly an unsustainable condition, inherited from half a century ago when the military was hungry for plutonium for bomb making. We are now faced with a sorcerer's apprentice grown wildly rogue, which we will only be able to control by a complete retreat from breeder technology.

Almost all of the SMRs now under development are designed to profoundly minimize this threat. All proposed designs produce small fractions of dangerous fissile material and several produce none at all. One design, from the laboratories of Duke University, proposes the use of an accelerator to transmute radioactive nuclei to stable isotopes in fission reactors, which, if developed would eliminate the present supply of weapons-usable material.

National Security

Although the need to restrain the emissions of CO_2 is paramount, there remain other insistant factors that demand attention and concern. At this writing, the security of the United States depends in part on the scale of the importation of oil from nations that are not in consonance with our way of life. It is not at all difficult to imagine a scenario such as the overthrow of the Saudi regime by militant Islamists more interested in a pan-Islamic hegemony than in piling up billions of Western dollars. In that scenario, previewed mildly in the 1973–74 OPEC embargo on oil to the United States, the predictable development would be once again to shut down the export of Saudi oil. This exposes the weakness of the U.S. energy policy, which remains dependent on uncontrollable external forces when dealing with the vastness of the oil volumes that flow westward every day.

Since there is simply no way to replace the Niagara of Middle Eastern oil with local production— we simply do not have the reserves—it becomes imperative to replace oil with nuclear energy on as short a timetable as possible. As has been argued, the present U.S. industrial timetable for conversion away from oil and other fossil fuels is conjured in decades, while disastrous changes in control of the flow of oil from the Middle East could take place tomorrow. This is, from the point of view of defense, an untenable national security profile that reminds us of how the terrifying threat of a Nazi atomic bomb drove the Manhattan Project. We are at no lesser level of threat from the fragility of Mid East politics than we were from the threat of Hitler's conquest of Europe. In fact, the threat is much more direct. The impact of the loss of Europe to Hitler, in the 1940s, while a moral tragedy, would have been much less economically damaging than would be the loss of Arab oil to America today.

Any national security policy must take into account that oil is more dangerous to our national existence than an atom bomb. Atom bombs are so 1980, while the flow of oil is the defining element in the history of the twenty-first century. While we are becoming alert to the threat of global warming, which lies in our more or less distant future, we have allowed ourselves to become blinded to the possible and predictable shutdown of Arab oil, which could be the headline in next week's newspaper.

John Kennedy got us to the moon in a mere handful of years. It took the unbridled commitment of the federal government to provide the push to get us out into space. The same sort of passion that would reflect the menace of too much oil too fast from sources that are too likely to be our enemies in the future could free us from gaps in our security profile in less than the time it took us to break the bonds of gravity.

The Fouling of America

National security aside, the problem of national pristinicity remains. This concern was heightened by the release by BP of more than 20 million barrels of heavy oil near the marshlands that surround New Orleans.

With uncounted oil wells devoted to producing a positive financial bottom line as well as oil, this disaster was inevitable, as will be the inevitable oil spills to come. The overwhelming quantities being handled are inherently uncontainable and are subject to natural and man-made accidents that are far beyond our ability to control. The scorching visuals of the blown wells fouling the atmosphere after the Iraqi invasion of Kuwait are magnified by the heartbreaking pictures of the devastation to all living things along our own Gulf Coast.

Oil is a foul and fouling raging beast and our ability to contain ever greater accidents is limited by tools demonstrably inadequate to the task. One natural method of cleanup proposed by a desperate oil industry proves to be a medication as bad as the disease. The reliance on natural bacteria to clean up an oil spill by "eating" it has the ironic consequence of adding directly to the production of released CO_2. In the eating scenario, trillions of bacteria take two atoms of precious oxygen out of the air, combine them with an atom of carbon, and produce a molecule of CO_2.

Our man-made tools designed to stop leaks are curiously nineteenth century. A mechanical "choke" that is intended to squeeze off the flow of escaping oil proved undependable. The hooding of a leak by a four-story steel bell jar did not work, and the suggestion that by simply dumping a load of junk (tires, bedsprings, golf balls, and so forth) leaks can be contained may be experimentally possible but is not likely to work in practice.

Oil is not the only culprit in the fouling of America; each of the fossil fuels adds to it. Coal contributes by blowing the tops off pristine mountains in wild places and choking us with coal dust and ash. Natural gas does its bit by fracking the rocky underpinnings of our land and fouling our water; We have yet to determine these consequences.

The Case for Miniaturization

As has been argued in this chapter, everywhere one looks in the energy industry there are problems that can be ameliorated and in some cases solved by merely scaling down the industry. Whether the problems are related to global warming or weapons-grade proliferation or costs of construction or lack of time or a score of other conditions, the answer seems clearly to be to make our industrial processes smaller, to demand less of our scarce resources, to minimize leakages and loss, and, in general, to follow in the footsteps of the electronics industry, which has created greater and greater miracles with less and less. So long as we remain subject to what Alvin Weinberg defined as the disease of gargantuism in the nuclear industry, problems of scale will always, as they have in the past, create moral, ethical, and industrial diseconomies that overwhelm any possible advantages of making things ever bigger.

The universal Bunsen burner of the carbon cycle that propels all life on the planet must not be tampered with, or we will have so badly fouled our nest that no redemption will be possible. We are about to evict ourselves from our own sweet home...and we have nowhere else to go.

As the departing dolphins said, "So long, and thanks for all the fish."

CHAPTER EIGHT

A Perfectly Reasonable Proposal

THERE ARE two factors central to the nuclear debate. The first is the futility of trying to reverse the flow of carbon by constructing slow-built, financially risky gigawatt reactors. The second is the resistance, by owners of and investors in gigawatt reactors, to the introduction of the more nimble mini, modular, safer nuclear technologies.

Attempts to speed up the construction of gigawatt reactors as well as attempts to introduce new technologies both come up against the operational realities of a free capitalist economy and would require wrenching and radical changes in how our economy mobilizes capital. One proposal would be to launch a program similar to that which landed us on the moon in only ten years. The goal of this proposal would be to reverse, in the coming decade, the existing ratio of electrical generation. The present ratio is 80 percent fossil fuels versus 20 percent carbon free sources.

Capitalist economies quite properly resist radical change as a threat to long-term invested capital. Consequently, to achieve quick change from fossil to nuclear we will require a new approach that does not violate conventional financial canon and does not raise the ire of invested capital. This approach requires a financial jujitsu move in which the strength of the system is used to prevail over its resistance.

It will be necessary to preserve the monetary capital invested in coal, oil, and natural gas while, paradoxically, eliminating the use and production of these fossil fuels. This could be accomplished by simply adjusting the rate of plant depreciation. By allowing mining and oil interests to completely depreciate their capital in substantially fewer years than they do now, perhaps even as few as one, these interests would accumulate a reservoir of capital that could naturally flow to the only industry that has the capacity to absorb the enormous sums that would be seeking investment. Indeed, if we were not in such a damned hurry because of the warming threat, a slow version of this process, inherent in a capitalist economy, would reduce the use of fossil fuels without need for adjustments to the rate of capital depreciation.

Since owners would now have windfall profits by continuing undepreciated production until the nuclear goal is fully reached, every ton of coal and every gallon of oil and every cubic meter of natural gas must bear a tax burden that would ultimately provide the financing for the conversion of fossil fuel to nuclear generation.

The numbers are interesting.

The United States mined 12 billion tons of coal this year. The average price at the mine head was $30 a ton. A $6 tax at the mine head would provide $360 billion over the ten-year life of the project, taking into account the declining use of coal during those years. The $6 could easily be passed along to consumers.

The United States consumes over 300 billion gallons of petroleum products each year. A 50-cent tax per gallon would raise another $750 billion.

The United States consumes around 700 billion cubic meters of natural gas each year. A 5-cent tax on each cubic meter would provide $35 billion.

Thus, over a trillion dollars could become available for painless conversion to nuclear electricity. The funds would be extracted from an industry that was being paid off by the shortened depreciation rate. At the present estimated cost of SMRs, thousands of these modular devices could be provided in a decade if assembly-line production could be rapidly developed, along with a speeding up of regulatory approval.

The trillion dollars would not have to be actually expended. This enormous fund could be used to guarantee capital invested in nuclear projects, which would allow normal private-capital markets to do their thing with lesser risk. There would still be considerable risk to investors in the opportunity cost to their capital in unsuccessful ventures, but their capital would be preserved.

Achieving assembly-line production[87] and fast regulatory approval would require an enormous jump-start in SMR research, which, to date, has enjoyed little support. Instead of the pittance of research money currently committed to SMR production, imagine what $10 billion or $20 billion could accomplish at that level.

This proposal would also entail the shifting of the approximately 10,000,000 jobs in fossil fuel use to nuclear. Considering that this is less than 7 percent of the U.S. workforce, this adjustment over ten years would be no challenge to the employment picture. Indeed, the 10,000,000 jobs represent a net loss of wealth to the nation due to the cost of the parallel importation of required levels of oil and gas. When this export of capital is stopped over the ten-year shift to nuclear, hundreds of billions of dollars would become available for solving employment and other transitional problems.

Under such a defined process, in which the best industrial and scientific minds would be brought together, the United States not only would solve its own long-term energy needs but would find itself the leading resource for SMRs for export to an international market hungry for electricity. One huge segment of this international market, the underserved third world, would be most quickly and economically energized by the low-cost and limber qualities of SMRs.

This perfectly reasonable proposal serves the interests of all parties without heaping up unmovable berms of resistance to change. It is designed to attend to many of our difficult long-term national concerns, such as fighting global warming, creating new, high-quality jobs, and searching out new export markets.

In ten years we could be awash in cheap and reliable electrical energy, having also created a vibrant new industrial sector. In the end we would emerge as the dominant supplier to the world of its most needed resource.

Coda

THE CENTRAL INVESTIGATION of this book has been directed at the scale of the nuclear industry. The book has argued that all anthropogenic challenges that put in question continued human existence on Earth are a matter of scale.

It was nature's unanticipated success with her human experiment, the evolutionary choice of brains over brawn, setting in motion the underlying scale problems that opened our Pandora's box of calamities. The history of man on Earth can best be viewed as a race between population and resources in which, for some millennia, population expansion leads and the Earth's resources have been straining to catch up. When population bloomed from 100 million brainy humans to a billion, the problems of scale emerged as the price we had to pay for success as a species.

The conversion of forests to agriculture, responding to the need to feed a burgeoning population, initiated the emerging problem of scale. The elimination of oxygen-emitting forests was mitigated to a large measure in the beginning of our population growth by the slow rate of change of the deforestation, which allowed an absorbable increase of CO_2 in the atmosphere. Natural processes, such as the ability of the oceans to take up CO_2, tamped down global warming. But as the scale of the release of warming gases exploded a few hundred years ago, our remaining forests and our seas, our first line of defense against CO_2 imbalance, could not cope and the level of CO_2 has risen alarmingly each year since 1800.

When human population climbed from a billion to six billion and these six billion reveled in the enormous energy content of coal, the scenario for disaster on a global scale came into play. The impact of the loss of forest paled in comparison to the havoc that the use of fossil fuels represented. In a world that was hungry for energy and, not incidentally, living on a Malthusian edge of food supply, coal burst upon us as manna from heaven. Coal was everywhere, easy to mine, and in enormous, almost unending supply. It generated the cheap heat needed to run the engines of early industrialization.

An unintended Faustian bargain was struck. The immediate cost of coal in the cities, dirt and pollution, were not out of sync with what urban man had lived with for centuries. It was beyond the science and the understanding of the time that burning vast millennial coal deposits would do little more than discommode the proximate few and benefit many. Again it was not the burning, it was the scale of the burning that dumped billions of tons of CO_2 into the atmosphere. We are now presented with a horrendous invoice that must be paid if we are to survive in anywhere near the comfort to which we have become accustomed.

It has been the intent of this book to argue that the scale of the warming catastrophe must be viewed primarily in terms of the continuing flow of CO_2 into the atmosphere. Every possible source of CO_2, no matter how small, must be identified and interdicted, since every fourth molecule of the gas will remain with us as a climate moderator for thousands of years. What we find is that all of the sources of energy including so-called green energy are CO_2-culpable and that each, in spite of claims to the contrary, adds its tiny mite or enormous mass to the climate changes looming in man's future.

The book argues that the scale of the consumption of fossil fuels is clearly unsustainable and, more to the point, that the feeble attempts to restrict CO_2 production are little more than a glossing over of the problem. Capping but not ending production of greenhouse gases only magnifies the unthinkable future costs of bringing the level of CO_2 and other greenhouse gases back into balance. Logic dictates that merely limiting greenhouse gases pushes possible solutions farther and farther into the future and does little to mitigate the difficulties that will arise in the near future. Logic dictates that our reasonably comfortable survival depends on the immediate and total cessation of increases to parts per million of CO_2 in the air. Logic dictates that if we are to continue to enjoy the level of comfort, wealth, and ease afforded us since the beginning of the twentieth century we must not only halt the increase but commence the actual decrease of warming gases at work in the atmosphere.

That conclusion brings the book to the problems and the solutions inherent in nuclear power, the only energy source that can guarantee us a reasonable future that might be resistant to CO_2 warming. Here the argument returns once again to the problem of scale of nuclear reactors, especially as the size of these reactors is related to the brief time left to us to get a grip on calamitous climate changes.

The beginnings of nuclear energy lay in the demands of war. The battle between good and evil characterized by the Second World War gave hurried birth to a discovery that had the inherent power to both destroy and salvage. The power to destroy required plutonium on an enormous scale, which was projected forward into the postwar development of civilian reactors. The demand for scarce plutonium for the bombs of the cold war defined the type of reactors that were being developed. These were the breeder reactors, which spewed out plutonium measured in tons that had previously been available only in ounces, and would continue to do so when the wartime need was far behind us. What was once precious, rare, and desirable has become dangerous nuclear waste, and the imperfectly perceived scale of the waste problem has seriously inhibited the logical growth and development of nuclear power.

By some unthinkable universal coincidence, nuclear power became available to man for war at the same time that it could prove to be the solution to man's greatest peacetime challenge. But the gigawatt nuclear power plants that emerged from the war had within them the seeds of their own severe limitation. The scale of the risks, real and imagined, grew exponentially as the scale of energy output grew only linearly. These risks, some merely perceived, some dangerously real and some financial, have conspired to restrict the enormous expansion of nuclear power that is needed to quickly replace our present consumption of energy from fossil fuels. The present rate of replacement of fossil with nuclear sources is at a pace that will have little impact on ultimately dealing with the CO_2 imbalance. This slow rate of change is compounded of public fears, bureaucratic regulatory mechanisms resistant to novel solutions, and a private capital market that is unable to conjure with the imagined and real risks of the huge gigawatt reactors that dominate the industry. It is a Gordian knot that cannot be unraveled but which can only be cut by a political sword that, alas, still lacks the edge to do the job.

By another rare act of cosmic fortuity, there is a parallel existing nuclear technology that, barring political interference, is capable of addressing the scale problems inherent in gigawatt reactors. From the beginning of the nuclear era, researchers such as Weinberg and Wigner and Teller developed small, inherently safe nuclear reactors that did not breed plutonium. This was reason enough for the military, balancing urgent demands on research and development budgets, to consign the concept of "smaller and safer is better" to dusty shelves in our national science attic.

This book has argued that small reactors that produce a tenth of the energy of the giants also generate inordinately less of the risk that inhibits growth of the industry. Construction of small reactors is a fraction of the cost of construction of gigawatt reactors. Thus the number of years that scarce capital is tied up and at risk is substantially reduced.

The book argues that a 100 MWe reactor[88] is a much bigger hardware bargain than a gigawatt reactor, which, from start to output, can cost $15 billion.

It is not only the hardware costs that contribute to the devilish details of risk. The problem is the inability of the market to accurately or even approximately estimate the real cost of the capital that would be tied up for over a decade in a project that, through technological advancements, could be obsolete before it ever joins the grid.

Underlying all of this concern about nuclear and about warming is the wider concern that pure scale is the true enemy. Ubiquitous in all human endeavors is the fact that as anything—a process, a device, an institution—gets larger, the field of maneuver gets smaller.

The larger the investment, the smaller the receptivity to change and the more likely that existing structures, even those that are not economical in the long term, will dominate the industry.

Ultimately, the mother of all the problems of scale is the inherent human drive to multiply that results in ever increasing demands on our earthly resources. It is the untrammeled scale of the growth of population that is the first mover. For all other industrial, agricultural, and societal problems there exist theoretical limitations on growth, yet there is no such mechanism when it comes to the exhortation to go forth and multiply. We simply continue to do it until there is no more room to lie down and give birth.

When we apply to population growth the reasonable concept that the larger the venture, the more constricting it is, we are faced with the appalling notion that the end of the process is further limitation of choice. It is not unlikely that the population will get so large and so hungry that there will be no room left for freedom. With that unnerving prospect in mind, our future may include the equally unthinkable condition of denial of reproduction.

Man's hegemony on Earth derives entirely from a lucky accident. When the dinosaurs and a myriad of other species went extinct due to climate change, a niche on the evolutionary ladder devolved onto a small horse like creature who eventually became Man. The price of Man's birth was the death of millions of species. The death of Man from another looming climate change will hardly move the needle of the history of our planet.

The dinosaurs were not given a second chance.
Neither will we.

APPENDIX A
Small Choices

For the lay observer, interested in making an informed opinion about the confusing claims and counterclaims being made by early proponents of small modular reactors, SMRs, the following are thumbnail sketches of the leading claimants. Here are excerpts from the releases of the developers themselves and paint a fluid picture of an industry in its earliest stages of development. Early developers have teamed up with industrial giants who are just beginning to recognize that the future of nuclear power might well be in 100 MWe rather than 1,000 MWe reactors.

Because of the stringent regulatory mind-set in the United States, it is likely that only one, or at most two, of the proposed small reactors will emerge as an accepted model and go forward into commercial production.

At this moment it is a horse race, with "favorite," "also ran," and "could do" competitors, so predictions are dangerous. But, also as in horse races, there is always a winner, however unlikely and however far back in the pack he, or she, started.

EM2 GENERAL ATOMICS

General Atomics Fast Reactor technology, EM2 (Energy Multiplier Module), has been around for decades but was shelved in the 1970s in the United States because of early cold war fears over nuclear proliferation. EM2 zeros in on that problem: what to do with nuclear waste. EM2 will use spent fuel rods as its power source, thus reducing total stockpiles of nuclear waste as it converts the 99 percent of the energy left unused in the rods. General Atomics estimates that depleted nuclear fuel in the United States contains the energy equivalent of 9 trillion barrels of oil— more than three times the known oil reserves in the world.

The reactor will be a compact cylinder, 60 by 16 feet, factory built, and capable of being delivered by semi, train, or plane. It is designed to produce 240 MW of electricity, a good match for processes that require high levels of heat, such as fertilizer or chemical production, as well as replacement of existing coal-fueled energy plants.

An important element would be the cooling by helium, thus eliminating the demands on ever more scarce water supplies and allowing siting in desert areas, where no water supplies exist.

Experts predict that perfecting the design, getting regulatory approvals, and building the first functioning EM2 reactor would take twelve years. It is conceivable, using assembly-line methods, that the second EM2 would roll off the line the next day.

GT-MHR GENERAL ATOMICS

The Gas Turbine–Modular Helium Reactor (GT-MHR) is a turbine-generating system powered by a passively safe nuclear reactor. It eliminates the need to make steam to produce electricity. The power plant includes one or more modular units in underground silos, each containing a reactor vessel and a power-production vessel.

Helium-cooled reactors can operate at much higher temperatures than today's conventional nuclear plants and produces 50 percent less high-level waste per KWh of electricity than current reactors. Additionally GT-MHR spent fuel offers a greater resistance to diversion and proliferation than LWR spent fuel. The plutonium content in GT-MHR waste is more than a factor of 20 lower than that for LWR waste.

The GT-MHR's decay heat will not cause a meltdown even if the coolant is lost. The reactor's low power density and geometry ensure that decay heat will be dissipated passively by conduction and radiation without ever reaching a temperature that can threaten the integrity of its ceramic-coated fuel particles, even under the most severe accident conditions.

HYPERION POWER MODULE

The Hyperion Power Module is a compact device capable of generating high levels of thermal power and is self-regulating to a constant temperature of operation. The thermal stability of the power module is built into the design of the nuclear reaction and is achieved without any mechanical moving parts or other external controls. The constant temperature characteristic allows the device to regulate its output in relation to how much power is drawn so that it can automatically accommodate power production. The absence of mechanical moving parts should make the reactor nearly maintenance-free for decades.

Buried 100 feet deep into the ground in a concrete box, it is self-contained, completely self-regulating, and small enough that it could be dug up at the end of its useful life, which is measured in years or decades, and taken back to a factory for dismantling and disposal. It addresses many of the major concerns that plague current reactor designs, such as high operating costs and extremely complex engineering safety controls.

This system achieves automatic "feedback" control through the use of hydrogen as a moderator. Radioactive metal hydride at the core has the natural property of absorbing and releasing hydrogen which acts to slow down radioactivity when the temperature rises and increase radioactivity as the temperature falls. Thus it is self-stabilizing and requires little or no active human control or monitoring. This reliance on natural law rather than on engineering safeguards makes the Hyperion model truly, inherently safe.

Concerning the latest Hyperion model adopted to accommodate the NRC, the Company reports that it is "sufficiently compact that it can be completely assembled, fuel and all, at the factory and then shipped sealed to the operating site. After about 10 years of operation the reactor core can be extracted from its secure underground vault and shipped back to the factory for storage or refurbishment and refueling. The sealed unit is not to be opened at any time after leaving the factory until it is returned to the factory. This eliminates any possibility of radioactive contamination at the operating site or any place along the transportation route.

This Hyperion power module uses a liquid metal coolant so that the reactor can operate at ambient pressure, eliminating the need for a pressure vessel. This is an important safety feature that distinguishes this reactor from the traditional light-water reactors that must contain steam pressures between 1500 and 2500 psi. In the Hyperion design, the steam boiler and turbine-generator are physically separated from the reactor and located above ground at a safe distance from the reactor. The liquid metal that was selected for this reactor is lead-bismuth that melts 23°C above the boiling point of water. The lead-bismuth eutectic does not react with either water or air, eliminating any threats of metal fires.

The new Hyperion reactor uses uranium nitride fuel instead of the commonplace uranium oxide. Uranium nitride is a robust, high-temperature ceramic that is chemically very inert. It has a thermal conductivity that is ten times higher that that of uranium oxide. That characteristic greatly reduces the temperature gradients that threaten the physical and chemical integrity of the traditional oxide fuels since their low thermal conductivity causes the centerline temperature in oxide fuel to rise hundreds of degrees over the coolant temperature. The large temperature difference induces structural stresses into the fuel, threatening to crack the fuel pellets. The high centerline temperature can cause the fuel chemistry to change, also weakening the oxide pellets. The nitride fuel does not have those problems."

MPOWER BABCOCK & WILCOX

B&W's mPower reactor would be used for smaller grids or limited electricity-demand areas, such as those of municipal districts or for individual industrial use and in developing countries whose transmission systems cannot handle large reactors.

The mPower reactor would include independent "modular" units that could be manufactured on an assembly line, thus cutting manufacturing and construction costs. Units could come on line even as others are being built, allowing power companies to start earning revenue faster. This brings lower installation base cost and, more important, greater cost certainty.

B&W plans to deploy the B&W mPower reactor — a scalable, modular, passively safe, advanced light water reactor system — which has the capacity to provide 125 MWe, for a five-year operating cycle without refueling, and is designed to produce clean, near-zero emission operations.

The reactors will offer several passive safety design features, including an underground containment that could accommodate storage for all of the spent fuel the reactor would use in a sixty-year operating life.

Passive safety systems
• Described as "Walk away safe."

Five-year operating cycle between refueling
• Somewhat shorter than competing systems

Five-percent enriched fuel
• This low level of enrichment precludes easy proliferation

Secure underground containment
• A self-evident advantage

Spent fuel pool capacity for the life of the reactor
• No need to retrieve and transport radioactive material for decades

Multiunit, modular installation
• Ten-plus units can be linked together for gigawatt output

North American shop-manufactured
• An enormous advantage compared to on site construction

Rail shippable
• Other systems ship directly to site by semi.

NUSCALE

The NuScale proposal is for a modular, scalable SMR version of the familiar light water reactor based on technology developed jointly by the DOE and Oregon State University. Essentially it is a miniature of the gigawatt plants that have already received regulatory approval. Delivery of the first unit from groundbreaking to site is estimated at thirty months. Subsequent units would be delivered on assembly-line schedules.

Dr. Jose Reyes, chief technology officer of NuScale, described a unit as a "stainless steel thermos, under water, underground." The firm has addressed safety issues throughout the design process: "Seismic isolators give remarkable seismic robustness," according to the CTO, and it is "walk-away safe" because of the water cooling design.

Each 45 MWe module has its own combined containment vessel and reactor system and its own designated turbine-generator set. Installations can be composed of just one or up to twenty-four units. Each unit can be taken out of service without affecting the operation of the others.

Each component is modular and is designed for factory fabrication. The containment and reactor vessel measures approximately 60 feet in length and 14 feet in diameter. All modular components are transportable by barge, truck, or rail. The reactor pressure vessel contains both the nuclear fuel the reactor and the steam generating system. Water in the reactor circulates using a convection process known as natural circulation. This is also described as a passive safety system because no pumps or other mechanical devices are required to circulate the water.

The design employs existing off-the-shelf technologies to minimize, and in many cases eliminate, the need for additional research and development. Since the primary coolant (water) is moved by natural circulation, the need for primary coolant pumps and external power is eliminated.

The reactor module, consisting of the containment and its contents, can be entirely fabricated at existing manufacturing facilities in the United States. As a result, construction can be on a significantly compressed schedule. NuScale plants will use nuclear fuel assemblies similar to those used in today's commercial nuclear plants. The only difference is the length of the fuel assemblies (6 feet for a NuScale system instead of the traditional 12 feet) and the reduced number of assemblies in the reactor. The NuScale proposal is just as its name implies, a traditional reactor design scaled down.

PEBBLE BED REACTOR

The fuel used in a Pebble Bed Reactor consists of spheres designed to contain radioactivity indefinitely.

Each sphere is made up of thousands of "particles." Each particle is covered with a special barrier coating, which ensures that radioactivity is kept locked inside the particle. One of the barriers, the silicon carbide barrier, is so dense that no gaseous or metallic radioactive products can escape. It retains its density up to temperatures of over 1,700 degrees Celsius, while the reactor design inherently limits temperatures to below 1,600 degrees. The reactor is loaded with almost half a million spheres, three-quarters of which are fuel spheres and a quarter graphite spheres. Fuel spheres are continually being added to the core from the top and removed from the bottom. The removed spheres are measured to see if all the uranium has been used. If it has, the sphere is sent to the spent fuel storage system, and if not it is reloaded in the core. An average fuel sphere will pass through the core about ten times before being discharged. The developers claim that used spheres will maintain their integrity for up to a million years, ensuring that spent fuel radioactivity is contained. Plutonium will have decayed away completely in 250,000 years.

The PBMR is "walk-away safe." Its safety is a result of the design, the materials used, and natural physical processes rather than engineered safety systems. This is because any increase in temperature makes the chain reaction less efficient, and it therefore ceases to generate power. The plant can never be hot enough for long enough to cause damage to the reactor.

PRISM (POWER REACTOR INNOVATIVE SMALL MODULE), GE AND HITACHI

PRISM is an integral fast reactor (IFR) being developed by GE with Hitachi.

It is designed for compact modular pool-type reactors with passive cooling for decay heat removal. Each unit consists of two modules of 311 MWe each, operating at high temperature (over 500°C). The modules are sited below ground and contain the complete generating system using a sodium coolant. The fuel is obtained from used light water reactor fuel. The amount of fuel that already exists for such reactors would be enough to power the world for millennia. Fuel stays in the reactor about six years, with one-third removed every two years. The system is designed to be installed in underground containment on seismic isolators with a passive air cooling ultimate heat sink. It is capable of modular installation with two reactor modules per turbine generator.

As a third-generation reactor, PRISM incorporates the passive, inherent safety features that require no active controls or operational intervention similar to other SMRs.

Reactor power: 840 MWt
Electrical output: 311 MWe
Outlet conditions: 930°F
Coolant: Liquid metal (sodium)
Refueling: 12–24 months

SSSS TOSHIBA

The Toshiba 4S (Super Safe, Small and Simple) is a "nuclear battery" reactor design. It requires only minimal staffing.

As envisioned, the Toshiba 4S would be able to supply about 10MW of electrical power for thirty years without any new fuel. It could be transported in modules by barge and installed in a building measuring 22 meters by 16 meters by 11 meters.

Depending on a variety of assumptions, the cost for power could range as low as 6 cents per kilowatt hour. The nuclear core heat source for this plant is quite compact; it is only about 0.7 meters in diameter and about 2 meters tall. This section of the plant would be at the bottom of the 30-meter-deep excavation inside a sealed cylinder, a location that provides an impressive level of nuclear material security. The active core material is a metallic alloy of uranium, plutonium, and zirconium.

The 4S uses neutron reflector panels around the perimeter to maintain neutron density. These reflector panels replace complicated control rods, yet have the instant ability to shut down the nuclear reaction in case of an emergency. Additionally, the Toshiba 4S utilizes liquid sodium as a coolant, allowing the reactor to operate 200 degrees hotter than if cooled by water. This means that the reactor is depressurized, as water turned to steam at this temperature would run at thousands of pounds per square inch. On paper, it has been determined that the reactor could run for thirty years without being refueled.

TWR (TRAVELING WAVE REACTOR), TERRAPOWER CORPORATION

TWR is a new type of nuclear reactor that, unlike other current reactor designs that use only enriched uranium for fuel, uses waste by-products of those enrichment process.

A small amount of 10 percent enriched uranium sparks the beginning of the process, but then the nuclear reactor runs on the waste product and can breed and consume its own fuel. The benefits are that the reactor doesn't have to be refueled or have its waste removed until the end of the life of the reactor (theoretically, a couple of hundred years). Using waste uranium reduces the amount of waste in the overall nuclear life cycle and extends the available supply of the world's uranium.

In a sense, the wave can be visualized as two waves: "a breeding wave moving just ahead of a burning wave that consumes the bred material. Visualize a cylinder a few meters long that contains U-238 or depleted uranium. A nugget of uranium enriched to 10 percent is put at one end of the cylinder and a wave 40 centimeters wide is built up that breeds and burns plutonium and produces heat as it propagates from one end to the other. It is claimed that it would take 50 to 60 years for the wave to go from one end to the other for a reasonable-sized core."[89]

THORIUM LFTR

The community that has kept small nuclear aficionados' hopes alive for decades has been the ad hoc Energy from Thorium site. In the race for support of industrial giants, the thorium process was nowhere to be found. It was only when it seemed that thorium, cheap and widely available almost everywhere, would not have an important seat at the table that Kirk Sorensen, an early and passionate supporter of the liquid thorium fluoride reactor (LFTR), announced he had been hired by Teledyne Industries as chief nuclear power researcher. Sorensen for years had been a proponent of thorium-based reactors, originally conceived in the 1950s and last researched in the United States in the early 1970s at Oak Ridge National Laboratory in Tennessee.

Teledyne, a giant among giants, has been reluctant to release details concerning its plans, but the combination of

Sorensen and Teledyne puts thorium right up there, if not with the present favorites, then certainly with the can-dos.

Advantages of the LFTR profile that have been discussed on the Energy from Thorium site for years were recently published by Robert Hargraves. They are an impressive argument for serious consideration of the concept.

1. With continuous refueling and continuous waste fission product removal, LFTR has no refueling outages.

2. It can change power output to satisfy demand, satisfying today's need for both base-load coal or nuclear power and expensive peak-load natural gas power.

3. LFTR operates at high temperature, for 50 percent thermal/electrical conversion efficiency, thus needing only half the cooling required by today's coal or nuclear plant cooling towers.

4. It is air-cooled, critical for arid regions of the western United States and many developing countries where water is scarce.

5. It has low capital costs: because of its compact heat exchanger, Brayton cycle turbine, and intrinsic safety features, and because cooling requirements are halved, it does not need massive pressure vessels or containment domes.

6. It will cost $200 million for a moderate-size 100 MWe unit, allowing incremental capital outlays, affordability to developing nations, and suitability for factory production, truck transport, and site assembly.

7. It will be factory-produced, lowering costs and lessening the time needed to build and enabling technological improvements to the system should they develop.

8. LFTR is intrinsically safe because overheating expands the fuel salt past criticality. LFTR fuel is not pressurized, and total loss of power or control will allow a freeze plug to melt, gravitationally draining all fuel salt into a dump tray. One hundred percent of LFTR's thorium fuel is burned, compared to 0.7 percent of uranium burned in today's nuclear reactors.

9. LFTR is proliferation-resistant, because LFTR U-233 fuel also contains U-232 decay products that emit strong gamma radiation, hazardous to any bomb builders who might somehow seize control of the power plant for the many months necessary to extract uranium.

10. In the LFTR, plutonium and other actinides remain in the salt until fissioned, unlike in today's solid fuel reactors, which must refuel long before these long-lived radiotoxic elements are consumed.

11. No plutonium or other fissile material is ever isolated or transported to or from the LFTR, except for importing spent nuclear fuel waste used to start the LFTR.

ADVANCED REACTOR CONCEPTS (ARC)

This reactor is immersed in low-pressure liquid sodium. The intermediate loop that carries the heat away is also liquid sodium. Transfer of heat to a turbine is by a suggested Brayton cycle turbine, which uses liquid CO_2 as the medium, yielding an expected 40 percent efficiency rate for heat transfer.

Fuel for the ARC-100 is sealed in the reactor, used for twenty years, and then returned to the factory for reprocessing. The reactor vessel is installed underground and is 15 meters high, with a diameter of about 7 meters. The fuel itself is enriched to an average of 14 percent.

HOLTEC INHERENTLY SAFE MODULAR UNDERGROUND REACTOR (HI-SMUR)

HI-SMUR reactor core is located completely underground, it is operated by gravity induced flow (no reactor coolant pump), it does not rely on off-site power for shutdown (*Inherently* Safe), and it can be installed as a single unit or a cluster at a site (*Modular*). HI-SMUR's principal safety credentials derive from locating the core underground in a reactor vessel that has no penetrations to provide a draindown path for the reactor coolant. Eliminating the reactor coolant pump and the need for emergency or off-site power to cool the reactor core in the event of a forced shutdown, are among the distinguishing design features of HI-SMUR. Other major features of HI-SMUR are its small footprint, minuscule site boundary dose, large inventory of coolant in the reactor vessel and its modularity, i.e., the freedom to build the number of units at a site to best suit the owner's projected power needs. The expected duration of the construction life cycle is 24 months.

APPENDIX B
Teller's Laws of Nucleonics

In 1996, Edward Teller and two associates proposed a new and more elegant design for the next generation of nuclear power plants to replace the touchy LWR designs that had defined the accepted profile of nuclear power in the cold war era. In his proposal, Teller laid down a set of "necessary design elements." These design elements had been the basis for extensive research by Alvin Weinberg and Teller himself in the 1950s. This early research had been sidetracked by the military's need for vast quantities of plutonium to power the thousands of warheads that were required to support Mutual Assured Destruction (MAD) the improbable foreign policy pursued by both the USSR and the United States. The plutonium came from the breeder reactors favored by the military. It was these proliferating breeder reactors that Teller and his group sought to replace.

Teller's recommendations for safety, control, waste, and proliferation could well be considered the Laws of Nucleonics, similar to Asimov's Three Laws of Robotics.[90] Both deal with the unintended consequences of new and highly complicated technologies and both have at their core the directive that new technologies shall inherently avoid harm to their inventors.

The nontechnical selections from the paper presented here are well worth the read.

Completely Automated Nuclear Reactors for Long-Term Operation

Edward Teller, Muriel Ishikawa, and Lowell Wood
(Stanford University, January 1996 [Abstract])

We discuss new types of nuclear fission reactors optimized for the generation of high-temperature heat for exceedingly safe, economic, and long-duration electricity production in large, long-lived central power stations. These reactors are quite different in design, implementation and operation from conventional light-water-cooled and -moderated reactors (LWRs) currently in widespread use, which were scaled-up from submarine nuclear propulsion reactors. They feature an inexpensive initial fuel loading which lasts the entire 30-year design life of the power-plant. The reactor contains a core comprised of a nuclear ignitor and a nuclear burn-wave propagating region comprised of natural thorium or uranium, a neutron reflector which also implements a thermostating function on the reactivity, a pressure shell for coolant transport purposes, and automatic emergency heat dumping means to obviate concerns regarding loss-of coolant accidents during the plant's operational and post-operational life.

These reactors are proposed to be situated in suitable environments at −100 meter depths underground, and their operation is completely automatic, with no moving parts and no human access during or after its operational lifetime, in order to avoid both error and misuse. The power plant's heat engine and electrical generator subsystems are located above-ground.

Introduction

One of the ancient needs of mankind regarding the physical environment has been for a "well" of high-grade heat, from which thermal energy could be conveniently drawn, whenever and to the extent desired. With the Industrial Revolution, this need became more focused, for heat engines, e.g., those in electrical generating plants, have become the prime movers of virtually all of modern civilization.

The recently-gained ability to release energy from neutron-induced fission of the nuclei of the actinide elements in principle seems to satisfy this need, for this energy form is remarkably dense and capable of being accessed at essentially any desired power level. Moreover, the actinide elements, though present in the Earth's crust at a mean density of only a few parts per million, have become readily and inexpensively available.

Public acceptance of nuclear fission reactors has been impeded by reactors which require installation of new fuel assemblies approximately 10 times over the lifetime of a typical power plant, with attendant dangers of accidents, complexity, cost and plant unavailability during refueling operations. The handling and disposal of the spent fuel assemblies has also proved problematic, though for mostly non-technical reasons

The management of nuclear fission heat production on timescales of minutes to hours has proved troublesome, due to significant fractions of nuclear fission energy appearing considerably after it has been requested, i.e., from beta decay of fission products; such phenomena raise the specter of loss-of-cooling accidents and of possible reactor core meltdown. Perhaps most seriously, severe mismanagement of nuclear reactors by their human operators can cause them to generate very high levels of thermal power over brief durations, which can lead to physical disruption of the reactor (possibly followed by dispersion of significant quantities of reactor products into the biosphere), as happened at Chernobyl. Even more seriously, reactor products may be diverted for military purposes.

It is important to realize that these are not physically required features of a nuclear fission chain reactor, but are merely the common characteristics of a particular class of reactors which descended from submarine propulsion reactors and have been nearly universally employed for civilian electricity generation. Indeed, the impeccable record of nuclear reactor safety in the U.S. naval nuclear propulsion program, which has demonstrated several thousand reactor-years of accident-free operation since its inception, contrasted with that of the disaster plagued Soviet naval reactor program, emphasizes that design, engineering and operational practices, not underlying physics, are actually the determinants of contemporary nuclear reactor safety and reliability.

At the same time, it would be remarkable if power reactors developed and optimized to reliably deliver a few tens of megawatts in a highly time-variable manner from an extremely compact configuration over a few months' submarine mission duration were also anywhere near optimal for central station thermal power generation, where pertinent time-scales are a few decades, power scales are in thousands of megawatts and economy and safety are the key figures-of-merit. It is a historic curiosity that nuclear power reactor design for central-station generation has evolved so little from its submarine power-plant antecedents. We therefore examine central-station nuclear power generation de novo, and inquire as to how the features of greatest value in this application area may specify improved reactor designs and operational practices.

Desired Design Features

In contrast to naval propulsion reactors, ones in central power stations may reasonably be extremely generous in their mass and volume budgets. In contrast to the exceedingly cost-tolerant military environment, however, economic considerations are of great importance in central power station construction and operation. Similarly, while reliability-of-service may be traded off to a significant extent for maximum performance in military circumstances, civilian electricity generation is somewhat less tolerant of "forced outages." The major requirement, however, is safety that is easily understood by the public and is not dependent on absence of human errors. Such considerations motivate the design of nuclear reactor heat sources for central station electricity generation which are literally nuclear heat wells: maintenance free sources of high-temperature heat-as-desired, hopefully created with little more cost or effort than that of digging a hole in the ground.

The average neutron multiplicity per fission is greater than 2 for u233 and Pu239, even when the fission is induced by neutrons bringing in zero kinetic energy. This raises the possibility of "breeding" (one neutron used to generate a readily fissionable nucleus and a second to cause its fission) of the actinides available in the Earth's crust. We have emphasized breeding reactor designs, for reasons of economy, efficient actinide mass utilization, reactor mass and volume minimization and, *quite importantly, total avoidance of reactor refueling*. In order to bring about long-term operation at a steady level, we use a propagating burn process—"breeding without reprocessing." This means a nuclear "burn wave" moving out from an initial small configuration enriched in fissile material into a far greater mass of fertile material. The leisurely character of beta decay—central to the breeding process in the actinide isotopes—offers a fundamental physical guarantee that explosive propagation of such nuclear burning is impossible. The reactions in the nuclear fuel will propagate with a characteristic speed of the order of a meter per year.

Indeed, we consider avoidance of reactor refueling to be a fundamental design driver. Furthermore, we demand that the reactor and all of the reactor products should be inaccessible after the reactor starts to operate. The associated advantages are cost avoidance, obviating the periodic outage of the power plant, personnel safety and utilization efficiency, resistance to diversion of reactor products to military use, and obviation of accident and inadvertences associated with human access to the reactor after its construction is completed, inspected and certified. We find compelling reasons neither in physics nor engineering for not loading a reactor with sufficient fuel, appropriately configured, to enable it to operate at full power for a central power-plant's entire lifetime. We specifically require such one-per-lifetime fuel-charging of the reactors which we consider. More specifically, we locate the reactor −100 meters underground in order to ensure safety of the biosphere in an obvious manner.

Simplicity of reactor construction is highly desirable, for reasons of overall economy, reliability and certifiability of construction and corresponding facility of reactor proliferation. We therefore emphasize reactor designs which consist of very little more than a pressure vessel for containing reactor coolant surrounding a shield reflector and the reactor's fuel charge. The monolithic pressure shell should be formed around its contents, with significant penetrations made only for coolant introduction and removal. All actuators and other mechanical devices should be required to operate only once, to commence initial reactor operation. At all later times, the reactor works without moving parts (except possibly at time of final shutdown).

We suggest that high-pressure helium should be used as the core coolant and that the electricity generation should occur on the Earth's surface. The reactor should have a strong negative temperature coefficient. Thus the removal of more heat from the reactor's core will accelerate the nuclear fission process. The reactor thus delivers energy on demand, and control rods become unnecessary.

Completely automated reactor operation is a requirement. Its great desirability is highlighted by the incident at Three Mile Island and the catastrophe at Chernobyl. Both of these reactor accidents were caused by human operator malfunction. We emphasize designs in which the reactor maintains a design core temperature within quite limited variation. Having power-regulating features in the design which throttle the nuclear reaction rate to correspond closely to the power extracted from the reactor appears to us to be feasible—and crucial to the prospect of eliminating human operator tampering with the reactor.

Nuclear power stations which have their never-refueled, never-disturbed reactors situated deep underground seem to be outstanding candidates for environmentally harmless electricity generation. Their ease of fabrication and operation may recommend them particularly strongly for rapid, low-risk, low-cost adoption as large-scale, long-lived heat sources for central power stations. *Their resistance to materials diversion for military purposes and their huge margin of intrinsic safety may make them politically quite attractive.*

We therefore expect rational allocation of resources to lead naturally first to exploration, then to development and finally to widespread proliferation of such reactors. In the Third World, the need for more electricity is apt to increase rapidly. It is important to satisfy this need, even in countries for which there is less than complete confidence in political stability. This puts particularly great emphasis on the availability of reactors for electricity generation which are difficult to misuse for military purposes and easy to operate without expert personnel.

Thus, the inexpensive, readily realized "nuclear heat wells" which we hope to develop may well prove to be an attractive option for large-scale electricity supply development in the early 21st century.

APPENDIX C
A Pessimistic Note

Global Warming Estimates Double in Severity According to New MIT Modeling
Jeremy Hance (mongabay.com, May 20, 2009)

Employing the MIT Integrated Global System Model, scientists have found that global warming could be more than twice as severe as previous estimates six years ago. The MIT Integrated Global Systems Model, which uses computer simulations to analyze the relationship between climatic changes and the global economy, found during 400 runs of the model that there is a 90 percent probability that temperatures will have risen 3.5 to 7.4 degrees Celsius by the end of the century.

The results "unfortunately largely summed up all in the same direction," says Ronald Prinn, the co-director of the Joint Program and director of MIT's Center for Global Change Science. "Overall, they stacked up so they caused more projected global warming."

Each of the 400 runs of the computer simulations incorporated slight variations that scientists selected as having an equal probability of turning out correct based on current data. The Integrated Global Systems Model differs from many other modeling systems since it incorporates data on human-activity—such as economic growth and energy use—as well as analyzing atmospheric, oceanic, and biological systems.

The model estimated that the median rate of surface warming by 2100 is 5.2 degrees, more than doubling the median rate of 2.4 degrees estimated six years ago. This difference is based on a variety of factors, including current economic data which shows a lower probability of society significantly decreasing greenhouse gas emissions than the more optimistic model six years ago.

In addition, the new modeling incorporated new data about cooling in the 20th century due to volcanic activity, occurrences which masked overall warming. The model also takes into account carbon emissions from soot, and, finally, the rate at which oceans are able to remove carbon dioxide from the atmosphere and transfer it to the ocean's depths has been lowered due to new studies.

Despite the scientists' gloomy findings, they found that strong policies to cut greenhouse gases proved as effective in curbing global warming as they did in past models.

Without action "there is significantly more risk than we previously estimated," Prinn says. "This increases the urgency for significant policy action. There's no way the world can or should take these risks."

Prinn notes that computer modeling is always based on current knowledge and so the accuracy of any modeling is dependent on the soundness of the researcher's available information.

"We do the research, and let the results fall where they may," Prinn explains. "We don't pretend we can do it accurately. Instead, we do these 400 runs and look at the spread of the odds."

According to Prinn, the odds look grim unless societies "start now and steadily transform the global energy system over the coming decades to low or zero greenhouse gas-emitting technologies."

APPENDIX D
An Optimistic Note

A Small Price for a Large Benefit
Robert H. Frank (February 20, 2010)

Forecasts involving climate change are highly uncertain, denialists assert—a point that climate researchers themselves readily concede. The denialists view the uncertainty as strengthening their case for inaction.

Organizers of the recent climate conference in Copenhagen sought, unsuccessfully, to forge agreements to limit global warming to 3.6 degrees Fahrenheit by the end of the century. But even such an increase would cause deadly harm. And far greater damage is likely if we do nothing.

The numbers—and there are many to choose from—paint a grim picture. According to recent estimates from the Integrated Global Systems Model at the Massachusetts Institute of Technology, the median forecast is for a climb of 9 degrees Fahrenheit by century's end, in the absence of effective countermeasures.

That forecast, however, may underestimate the increase. According to the same M.I.T. model, there is a 10 percent chance that the average global temperature will rise more than 12.4 degrees by 2100, and a 3 percent chance it will climb more than 14.4 degrees. Warming on that scale would be truly catastrophic.

Scientists say that even the 3.6-degree increase would spell widespread loss of life, so it's hardly alarmist to view the risk of inaction as frightening.

In contrast, the risk of taking action should frighten no one. Essentially, the risk is that if current estimates turn out to be wildly pessimistic, the money spent to curb greenhouse gases wouldn't have been needed to save the planet. And yet that money would still have prevented substantial damage. (The M.I.T. model estimates a zero probability of the temperature rising by less than 3.6 degrees by 2100.)

Moreover, taking action won't cost much. According to estimates by the Intergovernmental Panel on Climate Change a tax of $80 a metric ton on carbon dioxide—or a cap-and-trade system with similar charges—would stabilize temperatures by mid-century.

This figure was determined, however, before the arrival of more pessimistic estimates on the pace of global warming. So let's assume a tax of $300 a ton, just to be safe.

Under such a tax, the prices of goods would rise in proportion to their carbon footprints—in the case of gasoline, for example, by roughly $2.60 a gallon.

A sudden price increase of that magnitude could indeed be painful. But if phased in, it would cause much less harm. Facing steadily increasing fuel prices, for example, manufacturers would scramble to develop more efficient vehicles.

Even from the existing menu of vehicles, a family could trade in its Ford Explorer, getting 15 miles per gallon, for a 32-m.p.g. Ford Focus wagon, thereby escaping the effect of higher gasoline prices. Europeans, many of whom already pay $4 a gallon more than Americans do for gasoline, have adapted to their higher prices with little difficulty.

In short, the cost of preventing catastrophic climate change is astonishingly small, and it involves just a few simple changes in behavior.

The real problem with the estimates is that the outcome may be worse than expected. And that's the strongest possible argument for taking action. In a rational world, that should be an easy choice, but in this case we appear to be headed in the wrong direction.

This strange state of affairs may be rooted in human psychology. As the Harvard psychologist Daniel Gilbert put it in a 2006 op-ed article in *The Los Angeles Times*, "Global warming is bad, but it doesn't make us feel nauseated or angry or disgraced, and thus we don't feel compelled to rail against it as we do against other momentous threats to our species, such as flag burning."

People tend to have strong emotions about topics like food and sex, and to create their own moral rules around these emotions, he says. "Moral emotions are the brain's call to action," he wrote. "If climate change were caused by gay sex, or by the practice of eating kittens, millions of protesters would be massing in the streets."

But the human brain is remarkably flexible. Emotions matter, but so does logic. Even though we did not evolve under conditions that predisposed us to become indignant about climate change, we can learn to take such risks more seriously. But that won't happen without better political leadership.

Senator James Inhofe, a Republican from Oklahoma, has said that "the claim that global warming is caused by man-made emissions is simply untrue and not based on sound science." On compelling evidence, he's wrong. Yet he and his colleagues have the power to block legislation on greenhouse gases.

We don't know how much hotter the planet will become by 2100. But the fact that we face "only" a 10 percent chance of a catastrophic 12-degree climb surely does not argue for inaction. It calls for immediate, decisive steps.

Most people would pay a substantial share of their wealth—much more, certainly, than the modest cost of a carbon tax—to avoid having someone pull the trigger on a gun pointed at their head with one bullet and nine empty chambers. Yet that's the kind of risk that some people think we should take.

Robert H. Frank is an economist at the Johnson Graduate School of Management at Cornell University.

APPENDIX E
A Neutral Note

Cutting Greenhouse Gases Now Would Save World from Worst Global Warming Scenarios

Jeremy Hance (mongabay.com, April 14, 2009)

If nations worked together to produce large cuts in greenhouse gases, the world would be saved from global warming's worst-case-scenarios, according to a new study from the National Center for Atmospheric Research (NCAR).

The study found that, although temperatures are set to rise this century, cutting greenhouse gases by 70 percent the globe could avoid the most dangerous aspects of climate change, including a drastic rise in sea level, melting of the Arctic sea ice, and large-scale changes in precipitation. In addition such cuts would eventually allow the climate to stabilize by the end of the century rather than a continuous rise in temperatures.

"This research indicates that we can no longer avoid significant warming during this century," says Warren Washington, a scientist with NCAR and lead author of the study. "But, if the world were to implement this level of emission cuts, we could stabilize the threat of climate change and avoid catastrophe."

Using supercomputer studies with the NCAR-based Community Climate System Model, the researchers assumed that carbon dioxide levels could be maintained at 450 parts per million. The U.S. Climate Change Science Program has stated that 450 ppm is a reasonable target if nations quickly tackle climate change.

Holding carbon dioxide levels at 450 ppm would lead to global temperatures increasing another 0.6. degree Celsius (about 1 degree Fahrenheit); currently global temperatures have already risen 1 degree Celsius (nearly 1.8 degrees Fahrenheit). However, if business continues as usual, carbon dioxide levels are estimated to reach 750 ppm by the end of the century, leading to what many climatologists consider a catastrophic rise of 2.2 degrees Celsius (4 degrees Fahrenheit)—with no end in sight.

The models found that checking carbon dioxide levels at 450 ppm would keep sea levels from rising above 14 centimeters (5.5 inches) due to the thermal expansion of a warming ocean, though the model didn't factor in the amount of sea level rise that may occur due to melting glaciers. If carbon dioxide levels increase unchecked, sea level is expected to go up 22 centimeters (8.7 inches)— again not taking into account glacier melt.

Furthermore, Arctic ice in the summer would continue to melt about 25 percent, but stabilize by the end of the century if emissions are checked. Whereas, if not, the ice will shrink by three-quarters and continue disappearing. Under such a scenario some studies have predicted a complete loss of Arctic ice. In addition, cuts now will allow the Arctic ecosystem to survive largely intact, preserving birds, mammals, and fish, including saving an important fisheries industry.

Changes in precipitation, including drought in America's Southwest and increased precipitation in Canada, would be halved if emissions are cut as opposed to business as usual scenarios.

"Our goal is to provide policymakers with appropriate research so they can make informed decisions," Washington says. "This study provides some hope that we can avoid the worst impacts of climate change—if society can cut emissions substantially over the next several decades and continue major cuts through the century."

APPENDIX F
Rickover Comes Aboard

Admiral Rickover and the Cult of Personality
Dr. Paul R. Schratz

When Admiral Hyman G. Rickover cleared his desk and took final departure from the U.S. Navy and the Naval Reactors Branch on the last day of January 1982, it marked the end of an era. None of us can quite share the feeling, for no one else ever completed 63 years of continuous active service before heading for pasture at age 82. During the last hundred years, only a few names come to mind of those who have made a major impact on their navies or nation: Mahan, Fisher, Gorshkov. Rickover can join them. He changed the U.S. Navy's ship propulsion, quality control, personnel selection, and training and education, and has had far-reaching effects on the defense establishment and the civilian nuclear energy field.

The career of Hyman Rickover raises trenchant issues. Are we seeing the first of a new breed of technocratic flag and general officers? Is he the indispensable man whose personal drive created a nuclear navy by the force of his indomitable will over the backs of reluctant admirals refusing to be dragged into the twentieth century? What influences shaped the ruthless zeal of this wisp of a man driven to unsurpassed heights of excellence in building a nuclear navy?

From his entry into the U.S. Naval Academy in June 1918, Rickover was in conflict with the aristocratic WASP aura of Annapolis. (His family, living in a poor American Jewish neighborhood in Chicago, had come to America from Maków, Poland. Young Rickover, at age six, traveling steerage, lived off a barrel of salted herring except when

passengers threw oranges to him and other children looking up from the bowels of the ship.) Unfriendly and friend-
less, he soon learned to hate the Naval Academy.

Class standing is extremely important at the Naval
Academy. It determines relative seniority in the Navy at
graduation and the order in which one "makes his number"
for future promotions as vacancies occur; both pay and seniority are involved. Rickover stood far from the top of his
class. He was resented as a loner, a cutthroat with an abra-
sive personality, and he happened to be Jewish. Midshipmen
represent a cross section of the nation; any anti-Semitism
at the Academy reflects the nation at large. After all, three
of the seven Jews in Rickover's class rose to flag rank, a
percentage many, many times higher than normal.

In his naval service to mid-career, Rickover showed
little promise of future greatness. He volunteered for submarine duty but was not a particularly good submariner.
He rose to executive officer of the USS S-*48* but was not
selected for command. His pattern of sea and shore assignments up to the rank of captain was unimpressive. But in
the fall of 1946 he saw nuclear power "as an opportunity
for the Navy—and for himself." Rickover soon parlayed
the opportunity into a fortuitous dual responsibility to
the Atomic Energy Commission, later the Department of
Energy, and the Navy. Playing one against the other, he
used the exploding technology of nuclear power to project
his own career.

Proving himself a master at bureaucratic infighting,
he built his empire, sensing shrewdly what few others ever
realized, that congressmen prefer giving money to people
rather than institutions. Going before committees as an
individual and not as a Navy official, he gave a strong and
convincing impression that he spoke as a man of truth and
right, not to support the U.S. Navy but for his nuclear
navy. No witness attacked another flag officer or another
Navy program; Rickover could and did. He told congress-

men what they wanted to hear, but rarely heard from a government witness. "Those of us who have an objective, a desire to get something done, cannot possibly compromise and communicate all day long with people who wallow in bureaucracy, who worship rules and ancient routines." Thanks to outstanding preparation and delivery by a truly expert witness, his flawless performances generated their own fame in the press as a folksy, down-home philosopher.

Beneath the surface, however, was always the cold, unrelenting, ruthless workaholic, undermining the bureaucracy while creating his own. His was a management textbook for the inside operator, organized to the smallest detail, intolerant of error, devoting everything, including his personal life, to a cause, an obsession. Rickover established a constituency in Congress by superior salesmanship of his own product and skillful sowing of dissension and division in competing programs. He destroyed any competing nuclear program within his own organization and any person likely to emerge as a competitor or successor.

Admiral Rickover's greatest contribution was neither as a technician nor manager; his real genius lay in infusing into the Navy the pursuit of excellence, the genius of the insistence on taking infinite pains in the smallest detail in the development of nuclear energy. He set high standards of excellence as the norm and forced compliance.

In the nuclear program, Special Trust and Confidence, the traditional words on a U.S. military officer's commission, had no relevance. Everything and everyone was checked, rechecked, then checked again.

Only those with superior qualifications were considered for the nuclear program. Probing the minds and attitudes of potential officers and crewmen, the Rickover stamp reached every individual in the program. Over the years the famed interview became a legend. Ordeals of harassment, verbal abuse, banishment to a broom closet, demeaning indignities, and sometimes foul language—all sought to evaluate the individual under stress.

From the early USS *Nautilus* days, the nuclear program was marked more and more by the growing cult of the individual. Rickover created his own navy within the submarine navy. An essential element of the cult was to make himself the indispensable man both in the Navy and in the program. Rickover drove both the nuclear program and his own career, carrying his promotions beyond "the system" to the apex of four stars, then retention on active duty far beyond normal or even legal retirement limits. The bloody bureaucratic vendetta within the Bureau of Ships was only the background for his grand strategy: the nation must have an all nuclear navy; he would create and control it.

Having achieved brilliant success with the pressurized water coolant system in the *Nautilus* installation, innovation in other types of plants was stifled. The USS *Seawolf* plant, developed in tandem with *Nautilus,* utilized liquid sodium as coolant, promising much smaller and more compact reactors. Because of limitations in metallurgy, the system was unsuccessful. The program was scrapped, and its obvious superiorities were never again reexamined, even after twenty years of further progress in nuclear technology. Rickover put the kiss of death on programs generated within the Office of Naval Research and elsewhere for smaller, lightweight reactors that could reduce the enormous size and cost of nuclear-powered ships. None saw the light of day; all were thwarted as interference in his work. Truly it has been said, "The father of the last technological revolution is in the ideal position to stamp out the next one."

The cult of personality and the dominance of the Rickover program tended increasingly to isolate the nuclear Navy officers from the real Navy. Brilliant, carefully selected, and meticulously trained, they are superb engineers. The Rickover system trains engineers rather than professional naval officers. Through his insistence, the Naval Academy curriculum offers 80 percent of its courses in the hard sciences, 20 percent in the liberal arts. (West Point divides it 60–40; the Air Force Academy, 50–50.) Mid-career "nucs" were screened from war college assignments and from staff duties outside the narrow limits of their specializations as nuclear engineers. Shore duty of any sort is limited; of 1,500 billets in the Navy for nuclear specialists, only 122 are ashore.

Admiral Rickover made a great contribution to his country over an unsurpassed 63-year career of active service. Unlike most senior officers who retire in a blaze of ceremony and parades, he chose to pass up the traditional ritual and make his farewell on Capitol Hill. A grateful Congress struck a gold medal in his honor and, turning down President Reagan's invitation to serve as a consultant on civilian nuclear matters, this driven, lonely man, suddenly grown old and tired in the service, passed quietly from the scene.

APPENDIX G

The reports that follow have been submitted by groups that represent almost all acknowledged and informed scientists whose work relates to climate. It is the nature of the scientific method to resist categorical statements and to depend on heavily favored probabilities to make the case. This does not demonstrate a weakness of support for a proposition, it only recognizes that no matter how convincing present evidence is there exists improbable unforeseens and the equally improbable unknowables that must be acknowledged. The scientific method always advises caution and, in the light of that, it is remarkable how convinced of the fact of global warring the scientific community is.

in 2007 titled "Lighting the Way: Toward a Sustainable Energy Future."

Current patterns of energy resources and energy usage are proving detrimental to the long-term welfare of humanity. The integrity of essential natural systems is already at risk from climate change caused by the atmospheric emissions of greenhouse gases. Concerted efforts should be mounted for improving energy efficiency and reducing the carbon intensity of the world economy.

International Council of Academies of Engineering and Technological Sciences

In 2007, the International Council of Academies of Engineering and Technological Sciences issued a "Statement on Environment and Sustainable Growth":

> As reported by the Intergovernmental Panel on Climate Change (IPCC), most of the observed global warming since the mid-20th century is very likely due to human-produced emission of greenhouse gases and this warming will continue unabated if present anthropogenic emissions continue or, worse, expand without control.

CAETS, therefore, endorses the many recent calls to decrease and control greenhouse gas emissions to an acceptable level as quickly as possible.

Joint Science Academies' Statements

Since 2001, thirty-two national science academies have come together to issue joint declarations confirming anthropogenic global warming and urging the nations of the world to reduce emissions of greenhouse gases. The signatories of these statements have been the national science academies of Australia, Belgium, Brazil, Cameroon, Canada, the Caribbean, China, France, Germany, Ghana, India, Indonesia, Ireland, Italy, Japan, Kenya, Madagascar,

Malaysia, Mexico, New Zealand, Nigeria, Russia, Senegal, South Africa, Sudan, Sweden, Tanzania, Uganda, United Kingdom, United States, Zambia, and Zimbabwe.

2001

- Following the publication of the IPCC Third Assessment Report, sixteen national science academies issued a joint statement explicitly acknowledging the IPCC position as representing the scientific consensus on climate change science. The sixteen science academies that issued the statement were those of Australia, Belgium, Brazil, Canada, the Caribbean, China, France, Germany, India, Indonesia, Ireland, Italy, Malaysia, New Zealand, Sweden, and the United Kingdom.

2005

- The national science academies of the G8 nations, plus Brazil, China, and India, three of the largest emitters of greenhouse gases in the developing world, signed a statement on the global response to climate change. The statement stresses that the scientific understanding of climate change is now sufficiently clear to justify nations taking prompt action, and explicitly endorsed the IPCC consensus. The eleven signatories were the science academies of Brazil, Canada, China, France, Germany, India, Italy, Japan, Russia, the United Kingdom, and the United States.

2007

- In preparation for the Thirty-third G8 Summit, the national science academies of the G8+5 nations issued a declaration referencing the position of the 2005 joint science academies' statement and acknowledging the confirmation of their previous conclusion by recent research. "It is unequivocal that the climate is changing, and it is very likely that this is predominantly caused by the increasing human interference with the atmosphere. These changes will transform the environmental conditions on Earth unless counter-measures are taken." The thirteen signatories were the national science academies of

Brazil, Canada, China, France, Germany, India, Italy, Japan, Mexico, Russia, South Africa, the United Kingdom, and the United States.

2008

- In preparation for the Thirty-fourth G8 summit, the national science academies of the G8+5 nations issued a declaration reiterating the position of the 2005 joint science academies' statement and reaffirming that "climate change is happening and that anthropogenic warming is influencing many physical and biological systems." Among other actions, the declaration urges all nations to "take appropriate economic and policy measures to accelerate transition to a low carbon society."

2009

- In advance of the UNFCCC negotiations to be held in Copenhagen in December 2009, the national science academies of the G8+5 nations issued a joint statement declaring, "Climate change and sustainable energy supply are crucial challenges for the future of humanity. It is essential that world leaders agree on the emission reductions needed to combat negative consequences of anthropogenic climate change" and asserts that "climate change is happening even faster than previously estimated; global CO_2 emissions since 2000 have been higher than even the highest predictions, Arctic sea ice has been melting at rates much faster than predicted, and the rise in the sea level has become more rapid." The thirteen signatories were the same national science academies that issued the 2007 and 2008 joint statements.

Network of African Science Academies

In 2007, the Network of African Science Academies submitted a joint "statement on sustainability, energy efficiency, and climate change" to the leaders meeting at the G8 Summit in Heiligendamm, Germany:

A consensus, based on current evidence, now exists within the global scientific community that human activities are the main source of climate change and that the burning of fossil fuels is largely responsible for driving this change.

The IPCC should be congratulated for the contribution it has made to public understanding of the nexus that exists between energy, climate and sustainability.

Royal Society of New Zealand

Having signed on to the first joint science academies' statement in 2001, the Royal Society of New Zealand released a separate statement in 2008 in order to clear up "the controversy over climate change and its causes, and possible confusion among the public:"

The globe is warming because of increasing greenhouse gas emissions.

Measurements show that greenhouse gas concentrations in the atmosphere are well above levels seen for many thousands of years. Further global climate changes are predicted, with impacts expected to become more costly as time progresses. Reducing future impacts of climate change will require substantial reductions of greenhouse gas emissions.

Polish Academy of Sciences

In December 2007, the General Assembly of the Polish Academy of Sciences issued a statement endorsing the IPCC conclusions, and stated:

Problems of global warming, climate change, and their various negative impacts on human life and on the functioning of entire societies are one of the most dramatic challenges of modern times.

PAS General Assembly calls on the national scientific communities and the national government to actively support Polish participation in this important endeavor.

National Research Council (U.S.)

In 2001, the Committee on the Science of Climate Change of the National Research Council published "Climate Change Science: An Analysis of Some Key Questions." This report explicitly endorses the IPCC view of attribution of recent climate change as representing the view of the scientific community. Human-induced warming and associated sea level rises are expected to continue through the twenty-first century. The IPCC's conclusion that most of the observed warming of the past fifty years is likely to have been due to the increase in greenhouse gas concentrations accurately reflects the current thinking of the scientific community on this issue.

American Association for the Advancement of Science

In 2006, the American Association for the Advancement of Science adopted an official statement on climate change in which it was stated, "The scientific evidence is clear: global climate change caused by human activities is occurring now, and it is a growing threat to society....The pace of change and the evidence of harm have increased markedly over the last five years. The time to control greenhouse gas emissions is now."

European Science Foundation

In 2007, the European Science Foundation issued a position paper on climate change.

There is now convincing evidence that since the industrial revolution, human activities, resulting in increasing concentrations of greenhouse gases have become a major agent of climate change. The levels of greenhouse gases currently in the atmosphere, and their impact, are likely to persist for several decades. On-going and increased efforts to mitigate climate change through reduction in greenhouse gases are therefore crucial.

Federation of Australian Scientific and Technological Societies

Global climate change is real and measurable. Since the start of the 20th century, the global mean surface temperature of the Earth has increased by more than 0.7°C and the rate of warming has been largest in the last 30 years. Key vulnerabilities arising from climate change include water resources, food supply, health, coastal settlements, biodiversity and some key ecosystems such as coral reefs and alpine regions. As the atmospheric concentration of greenhouse gases increases, impacts become more severe and widespread. To reduce the global net economic, environmental and social losses in the face of these impacts, the policy objective must remain squarely focused on returning greenhouse gas concentrations to near pre-industrial levels through the reduction of emissions. The spatial and temporal fingerprint of warming can be traced to increasing greenhouse gas concentrations in the atmosphere, which are a direct result of burning fossil fuels, broad-scale deforestation and other human activity.

American Geophysical Union

The American Geophysical Union statement, adopted by the society in 2003 and revised in 2007, affirms that rising levels of greenhouse gases have caused and will continue to cause the global surface temperature to be warmer.

The Earth's climate is now clearly out of balance and is warming. Many components of the climate system— including the temperatures of the atmosphere, land and

ocean, the extent of sea ice and mountain glaciers, the sea
level, the distribution of precipitation, and the length of
seasons—are now changing at rates and in patterns that
are not natural and are best explained by the increased
atmospheric abundances of greenhouse gases and aerosols
generated by human activity during the 20th century.
Temperatures increased on average by about 0.6°C over the
period 1956–2006. As of 2006, eleven of the previous
twelve years were warmer than any others since 1850.
Recent changes in many physical and biological systems
are linked with regional climate change.

European Federation of Geologists

In 2008, the European Federation of Geologists issued the
position paper "Carbon Capture and Geological Storage:"

> The EFG subscribes to the major findings that climate
> change is happening, is predominantly caused by anthro-
> pogenic emissions of CO_2, and poses a significant threat to
> human civilization.

European Geosciences Union

In 2005, the divisions of atmospheric and climate sciences
of the European Geosciences Union (EGU) issued a position statement on global response to climate change. The
statement asserts that the IPCC "represents the state-of-
the-art of climate science supported by the major science
academies around the world and by the vast majority of
science researchers and investigators as documented by
the peer reviewed scientific literature."

Additionally, in 2008, the EGU issued a position statement on ocean acidification, which states, "Ocean acidification is already occurring today and will continue to intensify,
closely tracking atmospheric CO_2 increase. Given the potential threat there is an urgent need for immediate action."

Geological Society of America

In 2006, the Geological Society of America adopted a position statement on global climate change.

> The Geological Society of America (GSA) supports the scientific conclusions that Earth's climate is changing; the climate changes are due in part to human activities; and the probable consequences of the climate changes will be significant and blind to geopolitical boundaries. Furthermore, the potential implications of global climate change and the time scale over which such changes will likely occur require active, effective, long-term planning.

Geological Society of Australia

In July 2009, the Geological Society of Australia issued the position statement, "Greenhouse Gas Emissions and Climate Change."

> Human activities have increasing impact on Earth's environments. Of particular concern are the well-documented loading of carbon dioxide (CO_2) to the atmosphere, which has been linked unequivocally to burning of fossil fuels, and the corresponding increase in average global temperature. Risks associated with these large-scale perturbations of the Earth's fundamental life-support systems include rising sea level, harmful shifts in the acid balance of the oceans and long-term changes in local and regional climate and extreme weather events.

International Union of Geodesy and Geophysics

In July 2007, the IUGG adopted a resolution titled "The Urgency of Addressing Climate Change." In it, the IUGG concurs with the "comprehensive and widely accepted and endorsed scientific assessments carried out by the

Intergovernmental Panel on Climate Change and regional and national bodies, which have firmly established, on the basis of scientific evidence, that human activities are the primary cause of recent climate change." They state further that the "continuing reliance on combustion of fossil fuels as the world's primary source of energy will lead to much higher atmospheric concentrations of greenhouse gasses, which will, in turn, cause significant increases in surface temperature, sea level, ocean acidification, and their related consequences to the environment and society."

National Association of Geoscience Teachers

In July 2009, the NAGT adopted a position statement on climate change in which it asserts that "Earth's climate is changing [and] that present warming trends are largely the result of human activities."

> NAGT strongly supports and will work to promote education in the science of climate change, the causes and effects of current global warming, and the immediate need for policies and actions that reduce the emission of greenhouse gases.

Stratigraphy Commission of the Geological Society of London

In its position paper on global warming, the Stratigraphy Commission of the Geological Society of London declares:

> Global climate change is increasingly recognized as the key threat to the continued development—and even survival—of humanity. We find that the evidence for human-induced climate change is now persuasive, and the need for direct action compelling.

American Meteorological Society

The American Meteorological Society statement adopted by its council in 2003 said:

> There is now clear evidence that the mean annual temperature at the Earth's surface, averaged over the entire globe, has been increasing in the past 200 years. There is also clear evidence that the abundance of greenhouse gases in the atmosphere has increased over the same period. In the past decade, significant progress has been made toward a better understanding of the climate system and toward improved projections of long-term climate change.... Human activities have become a major source of environmental change. Of great urgency are the climate consequences of the increasing atmospheric abundance of greenhouse gases....Because greenhouse gases continue to increase, we are, in effect, conducting a global climate experiment, neither planned nor controlled, the results of which may present unprecedented challenges to our wisdom and foresight as well as have significant impacts on our natural and societal systems.

Australian Meteorological and Oceanographic Society

The society has issued a "Statement on Climate Change," wherein it concludes:

> Global climate change and global warming are real and observable....It is highly likely that those human activities that have increased the concentration of greenhouse gases in the atmosphere have been largely responsible for the observed warming since 1950. The warming associated with increases in greenhouse gases originating from human activity is called the enhanced greenhouse effect. The atmospheric concentration of carbon dioxide has increased by more than 30% since the start of the industrial age and is higher now than at any time in at least the past 650,000

years. This increase is a direct result of burning fossil fuels, broad-scale deforestation and other human activity.

Canadian Foundation for Climate and Atmospheric Sciences

In November 2005, CFCAS issued a letter to the prime minister of Canada stating:

> We endorse the conclusions of the IPCC assessment that "There is new and stronger evidence that most of the warming observed over the last 50 years is attributable to human activities."...There is increasingly unambiguous evidence of changing climate in Canada and around the world. There will be increasing impacts of climate change on Canada's natural ecosystems and on our socio-economic activities. Advances in climate science since the 2001 IPCC Assessment have provided more evidence supporting the need for action and development of a strategy for adaptation to projected changes.

Canadian Meteorological and Oceanographic Society

"CMOS endorses the process of periodic climate science assessment carried out by the Intergovernmental Panel on Climate Change and supports the conclusion, in its Third Assessment Report, which states that the balance of evidence suggests a discernible human influence on global climate."

Royal Meteorological Society (UK)

In February 2007, after the release of the IPCC's Fourth Assessment Report, the society issued an endorsement of the report. In addition to referring to the IPCC as the "world's best climate scientists," they stated that climate change is happening as "the result of emissions since industrialization and we have already set in motion the next 50 years of global warming—what we do from now on will determine how much worse it will get."

World Meteorological Organization

In its "Statement at the Twelfth Session of the Conference of the Parties to the U.N. Framework Convention on Climate Change," presented on November 15, 2006, the WMO confirms the need to "prevent dangerous anthropogenic interference with the climate system." The WMO concurs that "scientific assessments have increasingly reaffirmed that human activities are indeed changing the composition of the atmosphere, in particular through the burning of fossil fuels for energy production and transportation." The WMO further concurs that "the present atmospheric concentration of CO_2 was never exceeded over the past 420,000 years"; and that the IPCC "assessments provide the most authoritative, up-to-date scientific advice."

American Quaternary Association

The AMQUA has stated, "Few credible scientists now doubt that humans have influenced the documented rise of global temperatures since the Industrial Revolution," citing "the growing body of evidence that warming of the atmosphere, especially over the past 50 years, is directly impacted by human activity."

International Union for Quaternary Research

The statement on climate change issued by the INQUA reiterates the conclusions of the IPCC.

> Human activities are now causing atmospheric concentrations of greenhouse gasses—including carbon dioxide, methane, tropospheric ozone, and nitrous oxide—to rise well above pre-industrial levels....Increases in greenhouse gases are causing temperatures to rise....The scientific

understanding of climate change is now sufficiently clear to justify nations taking prompt action.

American Association of Wildlife Veterinarians

The AAWV has issued a position statement regarding "climate change, wildlife diseases, and wildlife health."

There is widespread scientific agreement that the world's climate is changing and that the weight of evidence demonstrates that anthropogenic factors have and will continue to contribute significantly to global warming and climate change. It is anticipated that continuing changes to the climate will have serious negative impacts on public, animal and ecosystem health due to extreme weather events. Furthermore, there is increasing recognition of the inter-relationships of human, domestic animal, wildlife, and ecosystem health as illustrated by the fact the majority of recent emerging diseases have a wildlife origin.

American Society for Microbiology

In 2003, the society issued a public policy report in which they recommend "reducing net anthropogenic CO_2 emissions to the atmosphere" and "minimizing anthropogenic disturbances of" atmospheric gases.

Carbon dioxide concentrations were relatively stable for the past 10,000 years but then began to increase rapidly about 150 years ago…as a result of fossil fuel consumption and land use change.

Of course, changes in atmospheric composition are but one component of global change, which also includes disturbances in the physical and chemical conditions of the oceans and land surface. Although global change has been a natural process throughout Earth's history, humans are responsible for substantially accelerating present-day changes. Outbreaks of a number of diseases, including Lyme disease, hantavirus infections, dengue fever, bubonic plague, and cholera, have been linked to climate change.

Australian Coral Reef Society

There is almost total consensus among experts that the earth's climate is changing as a result of the build-up of greenhouse gases. The IPCC (involving over 3,000 of the world's experts) has come out with clear conclusions as to the reality of this phenomenon. One does not have to look further than the collective academy of scientists worldwide to see the string [of] statements on this worrying change to the earth's atmosphere.

There is broad scientific consensus that coral reefs are heavily affected by the activities of man and there are significant global influences that can make reefs more vulnerable such as global warming....It is highly likely that coral bleaching has been exacerbated by global warming.

Institute of Biology (UK)

The UK's Institute of Biology states, "There is scientific agreement that the rapid global warming that has occurred in recent years is mostly anthropogenic, i.e., due to human activity." As a consequence of global warming, the institute warns that a "rise in sea levels due to melting of ice caps is expected to occur. Rises in temperature will have complex and frequently localized effects on weather, but an overall increase in extreme weather conditions and changes in precipitation patterns are probable, resulting in flooding and drought. The spread of tropical diseases is also expected."

Society of American Foresters

In 2008, the SAF issued two position statements pertaining to climate change in which it cites the IPCC and the UNFCCC.

Forests are shaped by climate....Changes in temperature and precipitation regimes therefore have the potential to dramatically affect forests nationwide. There is growing evidence that our climate is changing. The changes in temperature have been associated with increasing concentrations of atmospheric carbon dioxide (CO_2) and other GHGs in the atmosphere.

The Wildlife Society (international)

The Wildlife Society has issued a position statement titled "Global Climate Change and Wildlife."

Scientists throughout the world have concluded that climate research conducted in the past two decades definitively shows that rapid worldwide climate change occurred in the 20th century, and will likely continue to occur for decades to come. Few scientists question the role of humans in exacerbating recent climate change through the emission of greenhouse gases. The critical issue is no longer "if" climate change is occurring, but rather how to address its effects on wildlife.

The statement concludes with a call for "reduction in anthropogenic (human-caused) sources of carbon dioxide and other greenhouse gas emissions contributing to global climate change and the conservation of CO_2-consuming natural sinks."

American Academy of Pediatrics

In 2007, the academy issued the policy statement "Global Climate Change and Children's Health."

There is broad scientific consensus that Earth's climate is warming rapidly and at an accelerating rate. Human activities, primarily the burning of fossil fuels, are very likely (>90% probability) to be the main cause of this warming. Climate-sensitive changes in ecosystems are already being observed, and fundamental, potentially irreversible, ecological changes may occur in the coming decades. Conservative environmental estimates of the impact of climate changes that are already in process indicate that they will result in numerous health effects to children.

American College of Preventive Medicine

In 2006, the college issued a policy statement on "Abrupt Climate Change and Public Health Implications."

> The American College of Preventive Medicine (ACPM) accept the position that global warming and climate change is occurring, that there is potential for abrupt climate change, and that human practices that increase greenhouse gases exacerbate the problem, and that the public health consequences may be severe.

American Medical Association

In 2008, the AMA issued a policy statement on global climate change declaring that its members:

> support the findings of the latest Intergovernmental Panel on Climate Change report, which states that the Earth is undergoing adverse global climate change and that these changes will negatively affect public health.

The AMA supports educating the medical community on the potential adverse public health effects of global climate change, including topics such as population displacement, flooding, infectious and vector-borne diseases, and healthy water supplies.

American Public Health Association

In 2007, the association issued a policy statement titled "Addressing the Urgent Threat of Global Climate Change to Public Health and the Environment."

The long-term threat of global climate change to global health is extremely serious and the fourth IPCC report and other scientific literature demonstrate convincingly that anthropogenic GHG emissions are primarily responsible for this threat....U.S. policy makers should immediately take necessary steps to reduce U.S. emissions of GHGs, including carbon dioxide, to avert dangerous climate change.

Australian Medical Association

In 2004, the association issued the position statement "Climate Change and Human Health," in which it recommends policies "to mitigate the possible consequential health effects of climate change through improved energy efficiency, clean energy production and other emission reduction steps." This statement was revised again in 2008.

The world's climate—our life-support system—is being altered in ways that are likely to pose significant direct and indirect challenges to health. While "climate change" can be due to natural forces or human activity, there is now substantial evidence to indicate that human activity—and specifically increased greenhouse gas (GHGs) emissions— is a key factor in the pace and extent of global temperature increases.

Health will be affected by climate change as a result of extreme events such as storms, floods, heat waves, and fires and indirectly by longer-term changes, such as drought, changes to the food and water supply, resource conflicts, and population shifts.

Increases in average temperatures mean that alterations in the geographic range and seasonality of certain infections and diseases (including vector-borne diseases such as malaria, dengue fever, and Ross River virus and food-borne infections such as Salmonellosis) may result in the first detectable effects of climate change on human health.

Human health is ultimately dependent on the health of the planet and its ecosystem.

World Federation of Public Health Associations

In 2001, the group issued a policy resolution on global climate change.

> Noting the conclusions of the United Nation's Intergovernmental Panel on Climate Change (IPCC) and other climatologists that anthropogenic greenhouse gases, which contribute to global climate change, have substantially increased in atmospheric concentration beyond natural processes and have increased by 28 percent since the industrial revolution....Realizing that subsequent health effects from such perturbations in the climate system would likely include an increase in: heat-related mortality and morbidity; vector-borne infectious diseases...water borne diseases...[and] malnutrition from threatened agriculture.

World Health Organization

In 2008, the organization issued a report, "Protecting Health from Climate Change."

> There is now widespread agreement that the earth is warming, due to emissions of greenhouse gases caused by human activity. It is also clear that current trends in energy use, development, and population growth will lead to continuing—and more severe—climate change.

The changing climate will inevitably affect the basic requirements for maintaining health: clean air and water, sufficient food, and adequate shelter. Each year, about 800,000 people die from causes attributable to urban air pollution, 1.8 million from diarrhea resulting from lack of access to clean water supply, sanitation, and poor hygiene,

3.5 million from malnutrition, and approximately 60,000 in natural disasters. A warmer and more variable climate threatens to lead to higher levels of some air pollutants, increase transmission of diseases through unclean water and through contaminated food, compromise agricultural production in some of the least developed countries, and increase the hazards of extreme weather.

American Astronomical Society

The society has endorsed the AGU statement:

> In endorsing the "Human Impacts on Climate" statement [issued by the American Geophysical Union], the AAS recognizes the collective expertise of the AGU in scientific subfields central to assessing and understanding global change, and acknowledges the strength of agreement among our AGU colleagues that the global climate is changing and human activities are contributing to that change.

American Chemical Society

The society stated:

> Careful and comprehensive scientific assessments have clearly demonstrated that the Earth's climate system is changing rapidly in response to growing atmospheric burdens of greenhouse gases and absorbing aerosol particles (IPCC, 2007). There is very little room for doubt that observed climate trends are due to human activities. The threats are serious and action is urgently needed to mitigate the risks of climate change.

> The reality of global warming, its current serious and potentially disastrous impacts on Earth system properties, and the key role emissions from human activities play in driving these phenomena have been recognized by earlier versions of this ACS policy statement (ACS, 2004), by other major scientific societies, including the American Geophys-

ical Union (AGU, 2003), the American Meteorological Society (AMS, 2007) and the American Association for the Advancement of Science (AAAS, 2007), and by the U. S. National Academies and ten other leading national academies of science (NA, 2005).

American Institute of Physics

The governing board of the American Institute of Physics has endorsed a position statement on climate change adopted by the American Geophysical Union (AGU) Council in December 2003.

American Physical Society

In November 2007, the APS adopted an official statement on climate change.

> Emissions of greenhouse gases from human activities are changing the atmosphere in ways that affect the Earth's climate. Greenhouse gases include carbon dioxide as well as methane, nitrous oxide and other gases. They are emitted from fossil fuel combustion and a range of industrial and agricultural processes.

> The evidence is incontrovertible: Global warming is occurring. If no mitigating actions are taken, significant disruptions in the Earth's physical and ecological systems, social systems, security and human health are likely to occur. We must reduce emissions of greenhouse gases beginning now.

American Statistical Association

On November 30, 2007, the association's board of directors adopted a statement on climate change that endorses the IPCC conclusions.

EngineersAustralia (The Institution of Engineers Australia)

"EngineersAustralia believes that Australia must act swiftly and proactively in line with global expectations to address climate change as an economic, social and environmental risk....We believe that addressing the costs of atmospheric emissions will lead to increasing our competitive advantage by minimizing risks and creating new economic opportunities. EngineersAustralia believes the Australian Government should ratify the Kyoto Protocol."

International Association for Great Lakes Research

In February 2009, the association issued a fact sheet on climate change.

> While the Earth's climate has changed many times during the planet's history because of natural factors, including volcanic eruptions and changes in the Earth's orbit, never before have we observed the present rapid rise in temperature and carbon dioxide (CO_2).

> Human activities resulting from the industrial revolution have changed the chemical composition of the atmosphere....Deforestation is now the second largest contributor to global warming, after the burning of fossil fuels. These human activities have significantly increased the concentration of "greenhouse gases" in the atmosphere.

> As the Earth's climate warms, we are seeing many changes: stronger, more destructive hurricanes; heavier rainfall; more disastrous flooding; more areas of the world experiencing severe drought; and more heat waves.

American Geological Institute

In 1999, the AGI issued the position statement "Global Climate Change."

> The American Geological Institute (AGI) strongly supports education concerning the scientific evidence of past climate change, the potential for future climate change due to the current building of carbon dioxide and other greenhouse gases, and the policy options available.

Understanding the interactions between the solid Earth, the oceans, the biosphere, and the atmosphere both in the present and over time is critical for accurately analyzing and predicting global climate change due to natural processes and possible human influences.

American Institute of Professional Geologists

In 2009, the AIPG sent a statement to President Barack Obama and other U.S. government officials.

> The geological professionals in AIPG recognize that climate change is occurring and has the potential to yield catastrophic impacts if humanity is not prepared to address those impacts. It is also recognized that climate change will occur regardless of the cause. The sooner a defensible scientific understanding can be developed, the better equipped humanity will be to develop economically viable and technically effective methods to support the needs of society.

Statements by Dissenting Organizations

With the release of the revised statement by the American Association of Petroleum Geologists in 2007, no remaining scientific body of national or international standing is known to reject the basic findings of human influence on recent climate change.

Doran and Kendall Zimmerman, 2009

A poll performed by Peter Doran and Maggie Kendall Zimmerman at the University of Illinois received replies from 3,146 of the 10,257 polled Earth scientists. Results were analyzed globally and by specialization: 76 out of 79 climatologists who "listed climate science as their area of expertise and who also have published more than 50% of their recent peer-reviewed papers on the subject of climate change" believe that mean global temperatures have risen compared to pre-1800s levels, and 75 out of 77 believe that human activity is a significant factor in changing mean global temperatures. Among all respondents, 90 percent agreed that temperatures have risen compared to pre-1800 levels, and 82 percent agreed that humans significantly influence the global temperature. Petroleum geologists and meteorologists were among the biggest doubters, with only 47 percent and 64 percent, respectively, believing in human involvement. A summary from the survey states that:

> It seems that the debate on the authenticity of global warming and the role played by human activity is largely nonexistent among those who understand the nuances and scientific basis of long-term climate processes.

Statistics, 2007

In 2007, Harris Interactive surveyed 489 randomly selected scientists. Only 5 percent believe that human activity does not contribute to greenhouse warming; 84 percent believe global climate change poses a moderate to very great danger.

APPENDIX H
Nuclear Time Line
Source: U.S. Department of Energy

Pre-1940s

1895

- Wilhelm Röntgen discovers X-rays. The world immediately appreciates their medical potential. Within five years, for example, the British army is using a mobile X-ray unit to locate bullets and shrapnel in wounded soldiers in the Sudan.

1898

- Marie Curie discovers the radioactive elements radium and polonium.

1905

- Albert Einstein develops theory about the relationship of mass and energy.

1911

- Georg von Hevesy conceives the idea of using radioactive tracers. This idea is later applied to, among other things, medical diagnosis. Von Hevesy receives the Nobel Prize in 1943.

1927

- Herman Blumgart, a Boston physician, first uses radioactive tracers to diagnose heart disease.

DECEMBER 1938

- Two German scientists, Otto Hahn and Fritz Strassmann, demonstrate nuclear fission.

AUGUST 1939

- Albert Einstein sends a letter to President Roosevelt informing him of German atomic research and the potential

for a bomb. This letter prompts Roosevelt to form a special
committee to investigate the military implications of atomic
research.

DECEMBER 1941

- Japan bombs Pearl Harbor. The United States enters World
War II.

SEPTEMBER 1942

- The Manhattan Project is formed to secretly build the atomic
bomb before the Germans.

NOVEMBER 1942

- Los Alamos, New Mexico, is selected as the site for an atomic
bomb laboratory. Robert Oppenheimer is named the director.

DECEMBER 1942

- Fermi demonstrates the first self-sustaining nuclear chain
reaction in a lab under the squash court at the University of
Chicago. Soon after, a complex of top-secret nuclear production
and research facilities are built by the Manhattan Project
across the country.

1942–45

- The Clinton Engineer Works is built in Oak Ridge, Tennessee.
It is renamed the Oak Ridge National Laboratory after World
War II. The Clinton Pile, the first true plutonium production
reactor, begins operation in November 1943. By March 1945,
K-25 and other gaseous diffusion plants are in operation.

1943–45

- The Hanford Site is built in Richland, Washington, by the
Manhattan Project to produce plutonium. The first reactor
begins operation in September 1944.

FEBRUARY 1945

- Yalta Summit ratifies a divided postwar Europe.

APRIL 1945

• U.S. troops liberate Nazi concentration camp at Buchenwald.

MAY 1945

• Germany surrenders.

JULY 1945

• The United States explodes the first atomic device at a site near Alamogordo, New Mexico.

AUGUST 1945

• The United States drops atomic bombs on Hiroshima and Nagasaki. Japan surrenders.

MARCH 1946

• Winston Churchill proclaims an "iron curtain" has come down across Europe.

JULY 1946

• Atomic Energy Act (AEA) is passed, establishing the Atomic Energy Commission (AEC). The AEC replaces the Manhattan Project on December 31, 1946. The AEA places further development of nuclear technology under civilian (not military) control.

JULY 1946

• The United States tests a nuclear bomb on Bikini Atoll, an island in the Pacific. Four days later bikini swimsuit debuts at a French fashion show.

AUGUST 1946

• The Oak Ridge facility ships the first nuclear reactor–produced radioisotopes for civilian use to the Barnard Cancer Hospital in St. Louis. After World War II, Oak Ridge turns out numerous inexpensive radioactive compounds for medical diagnosis and treatment and for research and industrial applications.

APRIL – MAY 1948

- Nuclear tests in the South Pacific (Operation Sandstone) pave the way for mass production of weapons that previously had to be assembled by hand. By late 1948 the United States has fifty nuclear bombs.

JUNE 1948

- The Soviet Union begins the Berlin Blockade, cutting West Berlin off from the West. The United States begins vast airlift to keep Berlin supplied with food and fuel.

MAY 1949

- National Chinese forces led by Chiang Kai-shek retreat from mainland China to Formosa.

AUGUST 1949

- The Soviet Union detonates its first atomic device.

The 1950s

JANUARY 1950

- President Truman orders the Atomic Energy Commission to develop the hydrogen bomb (H-bomb).

FEBRUARY 1950

- Senator Joseph McCarthy launches a crusade to root out communism in America. "McCarthyism" is born.

JUNE 1950

- The Korean War begins as North Korean forces invade South Korea.

DECEMBER 1951

- The first usable electricity from nuclear fission is produced at the National Reactor Station, later called the Idaho National Engineering Laboratory.

OCTOBER 1952

• Operations begin at the Savannah River Plant in Aiken, South Carolina, with the start-up of the heavy water plant.

DECEMBER 1953

• In his "Atoms for Peace" speech, President Eisenhower proposes joint international cooperation to develop peaceful applications of nuclear energy.

JANUARY 1954

• U.S. Secretary of State John Foster Dulles announces U.S. policy of massive retaliation, that the United States would respond to any Communist aggression.
• The first nuclear submarine, USS *Nautilus*, is launched.

APRIL 1954

• Army-McCarthy hearings are on TV for five weeks. By the end, Senator McCarthy is publicly disgraced.

AUGUST 1954

• The Atomic Energy Act of 1954 is passed to promote the peaceful uses of nuclear energy through private enterprise and to implement President Eisenhower's Atoms for Peace program.

JULY 1955

• Arco, Idaho, becomes the first U.S. town to be powered by nuclear energy.

OCTOBER 1956

• Hungarian revolution is crushed by Soviet tanks.

NOVEMBER 1956

• Soviet premier Nikita Khrushchev tells the West, "History is on our side. We will bury you."

JULY 1957

• The Sodium Reactor Experiment in Santa Susana, California, generates the first power from a civilian nuclear reactor.

SEPTEMBER 1957

- The United States sets off first underground nuclear test in a mountain tunnel in the remote desert one hundred miles from Las Vegas.

OCTOBER 1957

- Radiation is released when the graphite core of the Windscale Nuclear Reactor in England catches fire.
- The Soviet Union launches Sputnik, the first spacecraft.
- The International Atomic Energy Agency (IAEA) is formed to promote the peaceful uses of nuclear energy and to provide international safeguards and an inspection system to ensure nuclear materials aren't diverted from peaceful to military uses.

DECEMBER 1957

- The first U.S. large-scale nuclear power plant begins operating in Shippingport, Pennsylvania.

OCTOBER 1959

- The Dresden-1 Nuclear Power Station in Illinois achieves a self-sustaining nuclear reaction. It's the first U.S. nuclear power plant built entirely without government funding.

The 1960s

- Civil rights movement picks up momentum, from lunch counter sit-ins in Charlotte and Greensboro in 1960, to the March on Washington in August 1963 led by Martin Luther King Jr., to the long, hot summer of 1967, when riots broke out in many major cities in the United States.
- Public feeling against the Vietnam War escalates throughout the decade.
- John Kennedy, Malcolm X, Robert Kennedy, and Martin Luther King Jr. are assassinated.
- Fourteen nuclear materials production reactors are operating during the decade. Fifteen commercial nuclear reactors are operating by 1969.

JUNE 1960

* Soviet premier Nikita Khrushchev pledges support for "wars of national liberation" in an address to the United Nations.

JANUARY 1961

* In his inauguration speech, President Kennedy says, "Let every nation know, whether it wishes us well or ill, that we shall pay any price, bear any burden, meet any hardship, support any friend, oppose any foe to assure the survival and success of liberty."

APRIL 1961

* Soviet Yuri Gagarin is the first man in space.
* Central Intelligence Agency–backed invasion of Cuba at the Bay of Pigs fails.

AUGUST 1961

* The Berlin Wall is erected between West Berlin and East Berlin.

SEPTEMBER 1961

* As part of a campaign to reduce the United States' vulnerability to nuclear attack, President Kennedy advises Americans to build fallout shelters. President Kennedy's letter in the September issue of *Life* magazine sets off a wave of "shelter-mania," which lasts for about a year.

OCTOBER 1962

* U.S. reconnaissance discovers Soviet missiles in Cuba. The United States blockades Cuba for thirteen days until the Soviet Union agrees to remove its missiles. The United States also agrees to remove its missiles from Turkey.

JUNE 1963

* The United States and Soviet Union set up a hotline (teletype) between the White House and the Kremlin.

- The United States cancels development of the neutron bomb, which theoretically would destroy life but leave buildings intact.

NOVEMBER 1978

- The Uranium Mill Tailings Radiation Control Act of 1978 directs the DOE to stabilize and control uranium mill tailings at inactive milling sites and vicinity properties. DOE forms the Uranium Mill Tailings Remedial Action (UMTRA) Program as a result.

MARCH 1979

- Three Mile Island nuclear power plant near Harrisburg, Pennsylvania, suffers a partial core meltdown. Minimal radioactive material is released.

JUNE 1979

- The United States and Soviet Union sign the second Strategic Arms Limitation Treaty (SALT II), which limits each side's arsenals and restricts weapons development and modernization.

NOVEMBER 1979

- American hostages are taken in Iran.

DECEMBER 1979

- The Soviet Union invades Afghanistan.

The 1980s

- Four nuclear materials production reactors are operating during the decade. One hundred eleven commercial reactors are operating in the United States by 1989.

OCTOBER 1980

- The West Valley Demonstration Project Act of 1980 directs the DOE to construct a high-level nuclear waste solidification demonstration at the West Valley Plant in New York. The only commercial nuclear fuel reprocessing plant in the United States, the West Valley Plant recovered uranium and plutonium from spent nuclear fuel from 1966–1972. Nearly 600,000 gallons of high-level nuclear waste are stored at the plant.

NOVEMBER 1980

• Single-shell nuclear waste storage tanks at the Hanford Plant in Washington no longer receive waste. The liquid waste is being transferred to newer design double-shell tanks.

DECEMBER 1980

• The Low-Level Radioactive Waste Policy Act is passed, making states responsible for the disposal of their own low-level nuclear waste, such as from hospitals and industry.
• The Comprehensive Environmental Response, Compensation, and Liability Act (also known as Superfund) is passed in response to the discovery in the late 1970s of a large number of abandoned, leaking hazardous waste dumps. Under Superfund, the Environmental Protection Agency identifies hazardous sites, takes appropriate action, and sees that the responsible party pays for the cleanup.

1982

• The Shippingport nuclear power plant, built in 1957, is retired. Congress assigns the decontamination and decommissioning of this commercial reactor to the DOE. This is the first complete decontamination and decommissioning of a reactor in the United States. The reactor vessel is shipped to a low-level waste disposal facility at Hanford, Washington. The site is cleaned and released for unrestricted use in November 1989.

JANUARY 1983

• The Nuclear Waste Policy Act of 1982 is signed, authorizing the development of a high-level nuclear waste repository.

MARCH 1983

• President Reagan terms the Soviet Union the "evil empire" and announces the Strategic Defense Initiative (Star Wars), a satellite-based defense system that would destroy incoming missiles and warheads in space.

NOVEMBER 1983

• DOE begins construction of the Defense Waste Processing Facility (DWPF) at the Savannah River Plant in South Carolina. DWPF will make high-level nuclear waste into a glass-like substance, which will then be shipped to a repository deep within the Earth for permanent disposal.

APRIL 1984

• In *LEAF (Legal Environmental Assistance Foundation) v. Hodel,* the court rules that the DOE's Y-12 plant in Tennessee is subject to the Resource Conservation and Recovery Act.

AUGUST 1985

• The Soviet Union announces a nuclear testing moratorium.

JANUARY 1986

• Soviet president Gorbachev calls for disarmament by the year 2000.

APRIL 1986

• Chernobyl Nuclear Reactor meltdown and fire occur in the Soviet Union. Massive quantities of radioactive material are released.

MARCH 1987

• President Gorbachev proposes elimination of European short- and medium-range missiles. Later, NATO and West Germany support Gorbachev's proposal, with some changes.

DECEMBER 1987

• President Gorbachev and President Reagan sign the Intermediate-Range Nuclear Forces (NIF) Treaty, the first arms treaty signed by the superpowers calling for elimination of a whole class of weapons—intermediate range missiles.

- Nuclear Waste Policy Amendments Act designates Yucca Mountain, Nevada, for scientific investigation as candidate site for the nation's first geological repository for high-level radioactive waste and spent nuclear fuel.

NOVEMBER 1989

- DOE changes its focus from nuclear materials production to one of environmental cleanup, openness to public input, and overall accountability by forming the Office of Environmental Restoration and Waste Management.
- The Berlin Wall is torn down. Many communist governments in Eastern Europe collapse.

1989

- Nuclear weapons production facilities at Rocky Flats Plant in Colorado and Fernald Feed Materials Production Center in Ohio cease production and change their missions to cleaning up their facilities.

The 1990s

OCTOBER 1990

- Germany is reunited as one country for the first time since the end of World War II.

NOVEMBER 1990

- Conference on Security and Cooperation in Europe formally ends the cold war and reduces Warsaw Pact and NATO conventional forces.

JULY 1991

- The United States and Soviet Union sign historic agreement to cut back long-range nuclear weapons by more than 30 percent over the next seven years.

1992

- The Hanford Site changes its mission from nuclear materials production to cleanup of its facilities.

OCTOBER 1992

- The Waste Isolation Pilot Plant (WIPP) Land Withdrawal Act withdraws public lands for WIPP, a test repository for transuranic nuclear waste located in a salt deposit deep under the desert.

DECEMBER 1992

- DOE's Office of Environmental Restoration and Waste Management (EM) and its predecessor agencies have decontaminated and dismantled more than ninety contaminated facilities across the country. EM has cleaned up eleven of forty-three sites under its Formerly Utilized Sites Remedial Action Program. Under its Uranium Mill Tailings Remedial Action Program, EM has cleaned up fifteen of twenty-four sites and 4,200 of 5,000 vicinity properties.

SEPTEMBER 1993

- Secretary of Energy O'Leary and Washington state's Governor Lowry host a two-day summit to make Hanford a model for the cleanup and revitalization of similar defense-related waste sites across the country.

1993...

- DOE continues to clean up the contamination from the past fifty years of the nuclear age. This contamination is the price we pay today for maintaining a strong national defense. DOE is working with regulatory agencies and the public to develop the technology needed for and to make the difficult choices associated with this national cleanup project.

APPENDIX I
Electric Vehicles

San Francisco Business Times (Jan. 29, 2010): As carmakers prepare to roll out a new generation of all-electric cars, the Bay Area's electricity system is far from prepared. By one assessment, the local electrical grid can handle no more than a few thousand electric cars charging up at peak hours—and that assumes the cars are distributed evenly at about one per block.

"It's the equivalent of adding another house to a transformer for a short period of time," said Cree Edwards, CEO of eMeter, which makes software to help utilities manage power. "The draw it has is significant."

To charge up its battery in just a few hours, an electric car needs to gulp power at a fast rate. The all-electric Tesla Roadster, for example, which can go more than 200 miles per charge, draws an amount of power "that could be as much as an entire house in the summertime at peak" in order to charge in 3½ hours, said Andrew Tang, senior director of PG&E's smart energy web division. Some Roadster owners have blown home fuse boxes and had to have wiring from their neighborhood transformer replaced, Tang said.

This could be especially troublesome in neighborhoods where multiple electric vehicle owners will live. Utilities have historically determined the size and capacity of electricity transformers, which distribute power to a group of users, based on modeling of expected peak usage. In residential areas, 10 to 16 homes are usually hooked to a single transformer. Transformers in commercial areas can handle larger loads.

Big, flat-screen TVs and more computers have made electricity use per household harder to predict because they are turned on and off and used at different times throughout the day.

That means many transformers are already near their limit at the busiest periods, which can extend through the day until 10 p.m. Two electric vehicles plugged into the same transformer when electricity use in the neighborhood is high could blow the transformer and black out the block, said Edwards and utility and electric car executives.

Increasing the size of transformers in neighborhoods is one solution PG&E is exploring for clustered electric vehicles. But with more than 1 million transformers in PG&E's service territories that becomes a huge and expensive problem

Utilities will have a better idea of the potential problem next year as the popularity of electric cars increases.

APPENDIX J
New York Railroad Storm

MAY 13, 1921: The New York Railroad Storm. The prelude to this particular storm began with a major sunspot sighted on the sun vast enough to be seen with the naked eye through smoked glass. The spot was 94,000 miles long and 21,000 miles wide and by May 14 was near the center of the sun in prime location to unleash an earth-directed flare. The flare, which was among the five worst events ever recorded, ended all communications traffic from the Atlantic Coast to the Mississippi. At 7:04 AM on May 15, the entire signal and switching system of the New York Central Railroad below 125th Street was put out of operation, followed by a fire in the control tower at 57th Street and Park Avenue. No one had ever heard of such a thing having happened during the course of an auroral display. The cause of the outage was later ascribed to a "ground current" that had invaded the electrical system. Railroad officials formally assigned blame to the aurora for a fire that destroyed the Central New England Railroad station. Telegraph Operator Hatch said that he was actually driven away from his telegraph instrument by a flame that enveloped his switchboard and ignited the entire building at a loss of $6,000, a significant sum in 1921. Overseas in Sweden a telephone station was "burned out," and the storm interfered with telephone, telegraph, and cable traffic over most of Europe. Aurora were visible in the Eastern United States, with additional reports from Pasadena, California.

APPENDIX K
Grid Disruption

"Of Solar Storms and the Electric Grid" (Sunday, February 7, 2010): From a *New Scientist* article discussing a NASA-funded study by the National Academy of Sciences of the effects of a severe solar storm on the electrical grid.

IT IS MIDNIGHT on 22 September 2012 and the skies above Manhattan are filled with a flickering curtain of colorful light. Few New Yorkers have seen the aurora this far south but their fascination is short-lived. Within a few seconds, electric bulbs dim and flicker, then become unusually bright for a fleeting moment. Then all the lights in the state go out. Within 90 seconds, the entire eastern half of the United States is without power.

A year later millions of Americans are dead and the nation's infrastructure lies in tatters. The World Bank declares America a developing nation. Europe, Scandinavia, China, and Japan are also struggling to recover from the same fateful event—a violent storm, 93 million miles away on the surface of the sun.

It sounds ridiculous. Surely the sun couldn't create so profound a disaster on Earth. Yet an extraordinary report funded by NASA and issued by the U.S. National Academy of Sciences (NAS) in January this year claims it could do just that.

Over the last few decades Western civilization has busily sown the many seeds of its own destruction. Our modern way of life, with its reliance on technology and especially electronics, has unwittingly exposed us to an extraordinary danger: plasma balls spewed from the surface of the sun that could wipe out our power grids, with catastrophic consequences.

The projections of just how catastrophic make chilling reading. "We're moving closer and closer to the edge of a possible disaster," says Daniel Baker, a space weather expert based at the University of Colorado in Boulder and chair of the NAS committee responsible for the report.

APPENDIX L
Nuclear Accidents

Los Alamos National Laboratory

Released by the Los Angeles National Laboratory,
Thomas McLaughlin, et. al.

LOS ALAMOS, N.M. (July 19, 2000): Since 1945 there have been 60 criticality accidents world-wide with varying levels of severity, from the most recent, a September 1999 accident in Japan that resulted in the deaths of two workers, to the very first fatal accident during the WWII Manhattan Project. All of these criticality accidents are now chronicled in a new report from the Department of Energy's Los Alamos National Laboratory, made public today.

The report, a joint effort of the United States and the Russian Federation, is the latest revision of a report first published in 1967 titled "A Review of Criticality Accidents." It includes significant new information about accidents that occurred in the United States, Russia, the United Kingdom, Canada, France, Argentina and Japan. The report begins by separating accidents into two distinct categories: process accidents and research reactor accidents.

Separating the two categories is important because of distinct differences between the two. In process accidents, administrative or physical controls are generally in place to prevent any sort of criticality, meaning that when criticality occurs it is wholly unwanted and unexpected. In research reactor accidents, some measure of criticality is purposely achieved, usually during experimentation, and ends up getting out of control somehow.

One of the report's principal authors, Thomas McLaughlin, of the Laboratory's Nuclear Criticality Safety group, says the report serves not only to reconstruct criticality accidents but to also offer information central to the prevention of accidents and the most effective response should an accident occur. "The most important sections of the report deal with observations and lessons learned from process criticality accidents," said McLaughlin. "All criticality accidents are dominated by design, managerial and operational failure, it's important that these issues be the focus for accident prevention."

A criticality accident occurs when the minimum amount of fissile material required to sustain a chain reaction is accidentally brought together. For example, when the nucleus of Uranium-235 disintegrates, two or three neutrons are released, and each is capable of causing another nucleus to disintegrate. However, if the total mass of the U-235 is insufficient to sustain a chain reaction, the neutrons simply escape. In most criticality accidents this chain reaction is very short lived, causing a neutron population spike and resultant radiation, heat and, in many cases, an ethereal "blue flash," a phenomenon of the air surrounding a neutron burst becoming ionized and giving off a flash of blue light.

The first-ever criticality accident resulting in a fatality occurred at Los Alamos' Omega Site on Aug. 21, 1945, and involved a 6.2-kilogram nickel-plated plutonium sphere and neutron-reflecting tungsten-carbide blocks. A critical assembly was created accidentally while the reflecting-blocks were stacked around the Pu sphere. The lone experimenter, Harry Daghlian, while moving a block in order to take a measurement had the block slip and drop onto the center of the assembly, causing a "superprompt" criticality event. Daghlian removed the dropped block by hand to stop the chain reaction and in doing so received a radiation dose estimated at 510 rem. He died 28 days later.

The vast majority of accidents do not result in death, but there have been 21 fatalities from criticality accidents since 1945. Nine of these fatalities were due to process accidents, and 12 resulted from research reactor accidents, according to the report. Of the 21 deaths, seven occurred in the United States, 10 in the Soviet Union, two in Japan, and one each in Yugoslavia and Argentina. There have been no accidents involving the transportation or storage of fissile materials. Only one accident, at the JCO Fuel Fabrication Plant in Tokimura, Japan, resulted in measurable exposures to members of the public, and these were well below allowable annual exposures to workers.

The report, originally scheduled for publication near the end of 1999, was held until now in order to add information from the process accident in Japan. "We felt it was very important to wait until we had the facts from this latest event so they could be added to this new revision," said McLaughlin. "The entire team working on this report wanted to make sure that it was a comprehensive as possible and contained the most up-to-date information."

In addition to McLaughlin, the report's authors are Shean Monahan of LANL, Norman Pruvost of Galaxy Computer Services, Inc., and Vladimir V. Frolov, Boris G. Ryazanov and Victor I. Sviridov all of the Institute of Physics and Power Engineering in Obninsk, Russia,

APPENDIX M

David Bushnell, NASA's chief scientist (July 5th, 2010):

THE PERMIAN EXTINCTION, which took place some 250 million years ago resulted in a decimation of animal life, leading many scientists to refer to it as the Great Dying. The Permian extinction is thought to have been caused by a sudden increase in CO_2 from Siberian volcanoes. The amount of CO_2 we're releasing into the atmosphere today, through human activity, is 100 times greater than what came out of those volcanoes.

During the Permian extinction, a number of chain-reaction events, or "positive feedbacks," resulted in oxygen-depleted oceans, enabling overgrowth of certain bacteria, producing copious amounts of hydrogen sulfide, making the atmosphere toxic, and decimating the ozone layer, all producing species die-off. The positive feedbacks not yet fully included in the IPCC projections include the release of the massive amounts of fossil methane, many times worse than CO_2 as an accelerator of warming, fossil CO_2 from the tundra and oceans, reduced oceanic CO_2 uptake due to higher temperatures, acidification and algae changes, changes in the earth's ability (Albedo) to reflect the sun's light back into space due to loss of glacier ice, changes in land use, and extensive water evaporation a greenhouse gas that occurs as a result of planetary temperature increases.

The additional effects of these feedbacks increase the projections from a 4°C–6°C temperature rise by 2100 to a 10°C–12°C rise, according to some estimates. At those temperatures, beyond 2100, essentially all the ice would melt and the ocean could rise by as much as 75 meters, flooding the homes of one-third of the global population.

Between now and then, ocean methane hydrate release could cause major tidal waves, and glacier melting could affect major rivers upon which a large percentage of the population depends. We'll see increases in flooding, storms, disease, droughts, species extinctions, ocean acidification, and a litany of other impacts, all as a consequence of man-made climate change. Arctic ice melting, CO_2 increases, and ocean warming are all occurring much faster than previous IPCC forecasts, so, as dire as the forecasts sound, they're actually conservative.

End Notes

CHAPTER 1

1. Not to slight Weinberg and Wiegand and Oppenheimer and Feynman and Szilard and a thousand others.

2. Chicago (IL): A team of international researchers has discovered that sulfuryl fluoride, a gas typically utilized in the repellant and control of insects, actually has an impact on global warming and is most definitely considered a greenhouse gas. In fact, though most are familiar with carbon dioxide (CO_2) the effect of sulfuryl fluoride on global warming is 4,800 times worse than that of CO_2.

3. Level as of 2010 at Mauna Loa Observatory, Hawaii.

4. Joseph Romm, "The Need for Speed: Hansen et al: We must phase-out coal emissions by 2030 and stabilize at or below 350 ppm" (www.grist.org/article/the-need-for-speed), November 2008.

5. Burkhard Bilger, "Hearth Surgery," *The New Yorker*, December 21, 2009.

6. Susan Solomon, Chemical Sciences Division, Earth System Research Laboratory, National Oceanic and Atmospheric Administration, Boulder, CO 80305.

Radiative forcing (RK) refers to an imbalance between incoming solar radiation and outgoing infrared radiation that causes the Earth's radiative balance to stray from its normal state. This balance is what is affected by the level of GHGs in the atmosphere.

Gian-Kasper Plattner, Institute of Biogeochemistry and Pollutant Dynamics, Zurich; Reto Knutti, Institute for Atmospheric and Climate Science, ETH CH-8092, Zurich, Switzerland; Pierre Friedlingstein, Institut Pierre Simon Laplace/Laboratoire des Sciences du Climat et de l'Environnement, Unité Mixte de Recherche 1572 Commissariat à l'Energie Atomique–Centre National de la Recherche Scientifique–Université Versailles Saint-Quentin, Commissariat a l'Energie Atomique-Saclay, l'Orme des Merisiers, 91191 Gif sur Yvette, France.

7. "Livestock's Long Shadow –Environmental Issues and Options," FAO, 2006

8. "Livestock and Climate Change," Worldwatch, December 2009.

CHAPTER 2

9. The idea of a safe reactor was originally conceived by Teller when his team of scientists assembled in the Little Red

Schoolhouse in San Diego in the summer of 1951. The mandate to this distinguished group, working under Teller, was to "design a reactor so safe…that if it was started from its shut-down condition and all its control rods instantaneously removed, it would settle down to a steady level of operation without melting any of its fuel." In other words, engineered safety was not good enough for Teller. The challenge was to design a reactor with inherent safety, guaranteed by the laws of nature.

10. World Nuclear Association.

11. The coal-fired unit, one of eight at TVA's Widow Creek Fossil Plant in North Alabama, broke the U.S. record in April 2005 for the longest continuous power production of any commercial power plant. This coal-fired plant operated continuously for 819 days.

CHAPTER 3

12. As we shall see, geothermal heat as it rises from the depths of the Earth usually carries with it greenhouse gases and other unpleasant chemicals.

13. The extraction from the seas of tidal energy which is powered by the moon, has yet to be demonstrated as a significant source of energy.

14. Wind and solar power have indirect deleterious effects on the environment, although both compare favorably to the use of fossil fuels.

15. Listed alphabetically with no relation to the CO_2 consequence of one method over another.

16. Royal Society and the Institution of Mechanical Engineers.

17. David Archer, *The Long Thaw*, Princeton University Press, 2009.

18. Almuth Ernsting and Rachel Smolker, "Biochar for Climate Change Mitigation: Fact of Fiction?" (www.biofuelwatch.org.2009).

19. U.S. Energy Information Administration International Energy Outlook, 2009.

20. Assume that high-quality coal is composed of 90 percent carbon, and that this carbon is released into the atmosphere as carbon dioxide when the coal is burned. The molecular weight of carbon is 12 and oxygen 16, so CO_2 weighs $12 + (2 \times 16) = 44$ mass units. Now multiply 90 percent of 100 tons of coal by 44/12, the ratio of the molecular weights of carbon dioxide to carbon. The result is 330 million tons of carbon dioxide, or 3.3 times the weight of the coal.

21. The typical coal train is 100 to 110 cars long, about a

mile of coal. Each hopper car holds 100 tons of coal, which lasts only 20 minutes fueling a power plant. Surface mines may load two or three trains of coal a day. In 1999, Wyoming alone shipped out 25,882 trains. That's almost 25,882 miles of coal—more than the circumference of the earth.

22. Power plants are the nation's biggest producer of toxic waste, surpassing industries such as plastic and paint manufacturing and chemical plants, according to a *New York Times* analysis of Environmental Protection Agency data. The EPA projects that by 2011 roughly 50 percent of coal-generated electricity in the United States will come from plants that use scrubbers or similar technologies, creating vast new sources of wastewater.

23. "Coal is coal and carbon has to go somewhere. Basically there is no such thing as 'clean coal.'" William Tucker, *Terrestrial Energy,* Savage, Maryland, Bartleby Press, 2008.

24. For each kilowatt-hour of electricity, about 2.1 pounds of CO_2, sent out as flue gas from the coal-fired power plants, are produced.

25. Jim Gardner, of Boise State University's Office of Energy Research, Policy and Campus Sustainability, notes that some additional energy (typically natural gas) is used during the expansion process to ensure that maximum energy is obtained from the compressed air. By some estimates, 1 kilowatt-hour's worth of natural gas will be needed for every 3 kilowatt-hours generated from a CAES system. "As is the case with any energy conversion, certain losses are inevitable," Gardner says. "Less energy eventually makes it to the grid if it passes through the CAES system than in a similar system without storage."

26. A flywheel is nothing more than a heavy wheel that rotates or spins freely. If you connect it the right kind of dual-purpose electric motor—some electric motors, like the ones in hybrid and electric cars, can function as both motors and generators—you can use the motor to spin the flywheel up to speed when there's a surplus of power. Then, when you need energy, you slow down the wheel by attaching it to a generator and convert its momentum back to electricity.

27. William F. Ruddiman, *Plows, Plagues and Petroleum,* Princeton University Press, 2005.

28. U.S. Department of Energy, 2007.

29. Based on Natural Gas Supply Association emission reports.

30. Seasonal changes in water depth mean there is a continuous supply of decaying material. In the dry season plants colonize the banks of the reservoir only to be engulfed when the water level rises. For shallow-shelving reservoirs these "drawdown" regions can account for several thousand square kilometers. In effect man-made reservoirs convert carbon dioxide in the atmosphere into methane. This is significant because methane's effect on global warming is 21 times stronger than carbon. National Maritime Museum UK, February 2005.

31. "The cold water pipe for that plant would be more than 33 feet in diameter—roughly the size of a metro tunnel in Washington, D.C.—and 3,000 feet long." *NDIA National Defense Industrial Association Bulletin*, April 2010.

32. On average, each hectare of grazing land supports no more than one head of cattle, whose carbon content is a fraction of a ton. In comparison, over 200 tons of carbon per hectare may be released within a short time after forest and other vegetation are cut, burned, or chewed. From the soil beneath, another 200 tons per hectare may be released. (Goodland and Anhang, Worldwatch, 2011.)

33. FAO report, *Livestock's Long Shadow—Environmental Issues and Options*, Henning Steinfeld et al.

34. Annual per capita meat consumption in developing countries doubled from 31 pounds in 1980 to 62 pounds in 2002, according to the Food and Agriculture Organization, which expects global meat production to more than double by 2050. That means the environmental damage of ranching would have to be cut in half just to keep emissions at their current, dangerous level.

35. Three explosions in natural gas–fired plants highlight the difficulties in controlling this fuel. An explosion in Connecticut in February 2010 killed five, an explosion in North Carolina in 2009 killed four, and an explosion in Michigan in 1999 killed six. In all cases, scores were injured. Compare fifteen natural gas plant deaths in only a few years to zero deaths in fifty years from commercial nuclear plants. The Connecticut explosion occurred during a "blowdown" procedure during which natural gas under high pressure is used to purge gas lines. The gas used is released directly into the atmosphere, adding to the CO_2 problem.

36. John Wright, CEO of Petrobank Energy & Resources.

37. An internal combustion engine is one in which the "fire" is contained within the mechanism producing energy. In the other kind of engine, steam, the "fire" burns, as in a boiler, outside the mechanism. In any event both are "burning."

38. "The Crude Truth About Oil Reserves," Leonardo Mauler, *Wall Street Journal,* November 4, 2009.

39. U.S. Energy Information Administration, International Energy Outlook, 2009.

40. In 2008, less than 1 percent of the world consumption of energy came from solar power. Even if a variation of Moore's law applies, solar power is not likely to be a serious help in solving planetary warming.

41. In a co-review of nuclear issues with *Energy in a Changing Climate*'s author Martin Nicholson, Professor Barry Brook, director of climate science at the University of Adelaide's Environment Institute, stated: "Renewable energy sources [such as wind and solar] use significantly more raw materials per unit of energy generated than even present-generation nuclear power stations and the full life-cycle emissions, including nuclear fuel production, are similar from both sources. When energy storage and fossil-fuel back-up are included, wind and solar emissions are much higher."

42. HBSC Traditional, onshore wind power breaks even with natural gas at $8.33 or oil at $92 a barrel. Offshore wind still needs a push: it requires natural gas at $17.14 or oil at $189 a barrel.Quick reality check: gas today is at $3 and oil is around $60.

43. David J. C. MacKay, *Sustainable Energy without the Hot Air*, Cambridge, England, UIT Publishers, 2009, p. 335, fig. 1.9.

44. Gavin Evans, "Electricity Prices on New Zealand's North Island Surged after Plant Maintenance and a Drop in Wind Generation Combined to Restrict Supplies," *Bloomberg News*, September 17, 2009.

45. Peter Lang, *Cost and Quantity of Greenhouse Gas Emissions Avoided by Wind Generation* (www.windaction.org/documents/20052).

46. U.S. Secretary of Energy Dr. Steven Chu.

47. Daniel Spreng, Gregg Marland, and Alvin M. Weinberg, "CO_2 Capture and Storage: Another Faustian Bargain?" ETH Zurich, CEPE—Centre for Energy Policy and Economics, 2007, Zürichbergstrasse 18, 8032

Zurich, Switzerland.

48. Phillip Boyd, *Nature Geosciences*, December 2009, volume 2, no. 12, pp. 809–98.

49. Adam Smith argues, in *The Wealth of Nations*, that each man operating in his own self-interest contributes to the satisfaction of all. Smith further stated, although it is generally ignored, that this happens only with the benign intervention of the "invisible hand of God." In other words, the personal pursuit of profit benefits society only when it is directed and controlled from outside the system.

CHAPTER 4

50. International Atomic Agency Commission.

51. "We have created a way to use fusion to relatively inexpensively destroy the waste from nuclear fission," says Mike Kotschenreuther, senior research scientist with the Institute for Fusion Studies (IFS) and Department of Physics. "Our waste destruction system, we believe, will allow nuclear power —a low carbon source of energy —to take its place in helping us combat global warming."

52. Section 216 allowed the Federal Emergency Regulatory Council (FERC) to make a decision concerning the siting, construction, or modification of electrical transmission lines if local authorities withheld approval for permit applications within one year. FERC proposed regulations concerning this authority. It concluded that the section not only allowed FERC to issue permits for projects where local permitting authorities failed to act on the applications within one year, but also allowed FERC to issue permits one year after local authorities denied a permit application. (Reported by Energy Practice Group, Strook & Strook & Lavan LLP, *Special Bulletin*, February 2010.)

53. "The vast system of electricity generation, transmission, and distribution that covers the United States and Canada is essentially a single machine—by many measures, the world's biggest machine." New policies "do not recognize the single-machine characteristics of the electric-power network," John Casazza wrote in 1998. "The new rule balkanized control over the single machine," he explains. "It is like having every player in an orchestra use their own tunes." ("The Industrial Physicist, What's wrong with the electric grid?" *Electrical World* 1998, 212 [4], 62–64.)

54. MLive.com.

55. In the most simple terms, a smart grid "talks back" to generating plants and allows supply to be adjusted by the generators to accommodate spikes and lows in demand. A smart grid gives the system the ability to diminish demand by shutting down noncritical use during peak hours and bring these uses back onto the grid when supplies are plentiful.

56. By one assessment, the local electric grid can handle no more than a few thousand electric cars charging up at peak hours—and that assumes the cars are distributed evenly at about one per block. (*San Francisco Business Times,* February 1, 2010.)

57. "684,000 electric vehicles will hit the streets within ten years. As motor manufacturers ramp up production of plug-in electric vehicles over the next few years, the report warns, these cars will require 3,785 megawatts of electrical output if charging simultaneously." ("Electric vehicles: largely dependent on coal, natural gas," John Guerrerio, examiner.com, March 24, 2010.)

58. "Every one of our nation's critical systems—water, healthcare, telecommunications, transportation, law enforcement, and financial services—depends on the grid." (Rep. Edward J. Markey, chairman of the Energy and Commerce Committee, reported in the *Boston Globe Green Blog,* March 24, 2010.)

59. Clifford D. May, Scripps Howard News Service, January 27, 2010.

60. *Severe Space Weather Events—Understanding Societal and Economic Impacts,* National Academy of Sciences, 2008.

61. Nuclear Electromagnetic Pulse, Jerry Emanuelson B.S.E.E., Futurescience LLC.

62. "Cyber security must address not only deliberate attacks, such as from disgruntled employees, industrial espionage, and terrorists, but also inadvertent compromises of the information infrastructure due to user errors, equipment failures, and natural disasters. Vulnerabilities might allow an attacker to penetrate a network, gain access to control software, and alter load conditions to destabilize the grid in unpredictable ways." *Smart Grid Cyber Security Strategy and Requirements,* National Institute of Standards and Technology, February 2010.

63. "The U.S. military is no more capable of operating without the Internet than Amazon.com would be," Richard

Clarke says. "Logistics, command and control, fleet positioning—everything down to targeting—all rely on software and other Internet-related technologies. The main reason for a ban on cyber war against civilian infrastructures is to defuse the current (silent but dangerous) situation in which nations are but a few keystrokes away from launching crippling attacks that could quickly escalate into a large-scale cyber war, or even a shooting war. The logic bombs in our grid, placed there in all likelihood by the Chinese military, and similar weapons the U.S. may have or may be about to place in other nations' networks, are as destabilizing as if secret agents had strapped explosives to transmission towers, transformers and generators." Richard Clarke was the former special adviser to the president for cyber security in 2001 and now teaches at Harvard's Kennedy School for Government.

64. And if all this were not enough to demonstrate how close to the edge we live, there are the endless tales, such as the following, that defy explanation. (Eric Wolff, wolf@nctimes.com.)

> The causes behind a brief but massive power failure last week that imperiled the Western electricity grid remained mysterious as of Friday evening, officials said. Shortly after midnight on Thursday, a drop in the supply of power generated inside San Diego County forced the California Independent Systems Operator, which manages the state's electric grid, to ask San Diego Gas & Electric Co. to cut power to 290,000 homes in order to prevent wider disruptions throughout the West.
>
> In the hours approaching midnight on the night of March 31, everything appeared to be under control. The Palomar Energy Center in Escondido and the South Bay power plant in Chula Vista were offline for scheduled maintenance. The San Onofre Nuclear Generation Station was operating at one-quarter power, but no one worried because the period between midnight and 4 a.m. is the part of the day with the lowest consumption.
>
> At midnight, the Otay Mesa Energy Center, a 600-megawatt power plant, powered down, as had been scheduled by the ISO, the nonprofit responsible for balancing electrical needs for the whole state. Not long after the plant was shut down, alerts started to go off at the ISO operations center in Folsom. The percentage of power produced inside San Diego County had dropped below the required 25 percent, and there wasn't enough transmission capacity available to bring in more from the outside. If no one acted, the imbalance in the grid could cause power disruptions *all the way up the west coast* (italics mine).

279

ISO operators called engineers at SDG&E and told them to reduce the load on the system. The utility cut power to some 290,000 homes and businesses scattered throughout San Diego County and southern Orange County. Within an hour, additional power sources had been brought online, and the crisis passed.

65. Based on Department of Energy (DOE) estimates.

66. In 2007 the Research and Innovative Technology Administration (RITA) reported that there were 1,520,200 miles of gas pipeline. in the United States. U.S. Department of Transportation (US-DOT).

CHAPTER 5

67. "Nuclear catastrophe was hanging by a thread...and we weren't counting days or hours, but minutes." —Soviet general and army chief of operations Anatoly Gribkov.

In 1962, the Soviet Union planted nuclear weapons ninety miles off the coast of Florida. The USSR lacked intercontinental ballistic missiles, while the United States held the advantage of being able to strike Russia from bases in Europe. Since we had sought bases close to their borders, the Russians felt that it was only proper that they have the same capability, thus correcting a perceived imbalance in MAD.

But the United States, never having had a direct continental threat on its borders, exhibited an unexpectedly paranoid reaction. Eventually the Russians, well understanding paranoia from their own national experience, backed down, with mere minutes to spare.

Ever more heated exchanges of threat and counterthreat ensued. The public, terrorized by huge black headlines, flights of fiction, and wild rumors, were digging bomb shelters in basements as their kids ducked under desks that would not protect them from a flyswatter.

Others, closer to the truth, were, as one describes it, "traumatized beyond terror." Some Kennedy Administration officials hopelessly watching the march toward Armageddon and, knowing how close we were to the end of human society, simply withdrew into a fog of apathy. One government official, a father known for his workaholic devotion to a twenty-hour-a-day high-security job, started coming home early, gathering his kids around him, and listening bleakly to news reports that he knew, bad as they sounded, were a pale whitewash of what was to come. There were thousands like him.

Looking back, the historical impact of those unthinkable days comes into sharp focus in the antinuclear movement of the four decades that followed. The nerve-shattering impact of the Cuban missile crisis, enhanced by powerful books and films, suggests that it is no wonder a national antinuclear panic developed; it is more to wonder that the panic is finally beginning to give way to a public acceptance of nuclear generated energy.

CHAPTER 6

68. May 2008: For two new AP1000 reactors at the Virgil C. Summer Nuclear Generating Station in South Carolina, South Carolina Electric and Gas Co. and Santee Cooper expected to pay $9.8 billion.

November 2008: For two new AP1000 reactors at its Lee site Duke Energy Carolinas raised the cost estimate to $11 billion, excluding finance and inflation, but apparently including other owners' costs.

November 2008: For two new AP1000 reactors at its Bellefonte site TVA updated its estimates for overnight capital cost esti-

mates to a total of $9.9 billion to $17.5 billion.

April 2008: Georgia Power Company reached a contract agreement for two AP1000 reactors to be built at an estimated final cost of $14 billion plus $3 billion for necessary transmission upgrades. (Wikipedia)

69. In some countries (notably the United States), in the past unexpected changes in licensing, inspection, and certification of nuclear power plants added delays and increased construction costs.

70. The Blue Castle project would require about 70 cubic feet per second for cooling water, or about 50,000 acre-feet of water per year. During a low-flow year, such as the one the Green River experienced in 2002, the plant would require about 2 percent of the river's flow through the town. (Gary Harmon, *the Grand Junction* [*Colorado*] *Daily Sentinel.*)

71. Between the years 1000 and 2000, the population of the world grew from approximately 300 million to 6 billion. Projections by Population Action International and other agencies suggest a world population in excess of 9 billion by the year 2030.

CHAPTER 7

72. Loss numbers vary between 7 percent and 12 percent.

73. For an in-depth discussion of this problem, see Smart Grid Cyber Security Strategy and Requirements, National Institute of Standards and Technology (NIST), U.S. Department of Commerce, 2010.

74. Clarke, who teaches at the Kennedy School of Government at Harvard, argues that smart grids will be even less secure against cyber attack than our present "simple" grids largely because a modernized U.S. power grid will be more highly networked and automated, with more points of attack.

75. There would inevitably be some local resistance losses, but they would be less than those from a national grid, with its thousand-mile reach.

76. Babcock & Wilcox's latest cost estimate of a prototype 125 MWe plant proposed to supply all of the electricity needs of the Oak Ridge National Laboratory. Since this prototype would bear the burden of the developmental costs leading up to it, we can reasonably expect that the subsequent price for plants would be some fraction of that.

77. Subassemblies of, for example, the engines, instrumen-

tation, and electronic, hydraulic, and other systems account for the seven-month figure.

78. Their size would also increase flexibility for utilities, since they could add units as demand changes, or use them for on-site replacement of aging fossil fuel plants. Some of the designs for SMRs use little or no water for cooling, which would reduce their environmental impact. Finally, some advanced concepts could potentially burn used fuel or nuclear waste, eliminating the plutonium that critics say could be used for nuclear weapons. (Steven Chu, U.S. Secretary of Energy.)

79. On the other hand, to be fair, we might well have a bit less CO_2 in the atmosphere.

80. During World War II the Boeing aircraft plant in Seattle turned out sixteen complete B-17s a day.

81. Potential impacts of a warming climate on water availability in snow-dominated regions, T. P. Barnett, J. C. Adam, D. P. Lettenmaier, *Nature* 438, 303–9 (November 17, 2005.)

82.

- 97.5 percent of all water on Earth is saltwater.

- Nearly 70 percent of the 2.5 percent of fresh water is frozen in the icecaps of Antarctica and Greenland; most of the remainder is present as soil moisture or lies in deep underground aquifers as groundwater, not accessible to human use.

- Less than 0.007 percent of all water on Earth is accessible for direct human uses. This is the water found in lakes, rivers, reservoirs, and those underground sources that are shallow enough to be tapped at an affordable cost. (www.GlobalChange.com, Human Appropriation of the World's Fresh Water Supply, accessed January 2006.)

83. Estimates from makers of small nuclear plants suggest that twenty 100 MWe plants could be sited in three years at a third of the cost of one 1,000 MWe plant, assuming that the regulatory blockades could be dealt with.

84. "Each year humanity dumps roughly 8 billion metric tons of carbon into the atmosphere, 6.5 billion tons from fossil fuels and 1.5 billion from deforestation." (*National Geographic*, February 2004.)

85.

- Advanced Boiling Water Reactor design by GE Nuclear Energy, last updated May 1997.
- System 80+ design by Westinghouse, last updated May 1997.
- AP600 design by Westinghouse, last updated December 1999.
- AP1000 design by Westinghouse, last updated January 2006.

86.

- International Reactor Innovative and Secure (IRIS), Westinghouse Electric Company.
- NuScale, NuScale Power, Inc.
- Pebble Bed Modular Reactor (PBMR), PBMR (Pty.), Ltd.
- Super Safe, Small and Simple (4S), Toshiba Corporation.
- Hyperion Power Generation, Inc.
- Power Reactor Innovative Small Module (PRISM), GE Hitachi Nuclear Energy.
- mPower, Babcock & Wilcox Company.

CHAPTER 8

87. In 1940 the United States was producing 2,000 stick-built military aircraft a year. Only six years later, responding to the demands of a wartime economy, the country had built an assembly line–based production of 303,000 military aircraft a year: an increase by a factor of 150. The U.S. population at the time was 139,000,000, with an industrial workforce of about 60,000,000, further diminished by 10,000,000 to 15,000,000 in military service. With the U.S. population in 2010 around 300,000,000 and a workforce of 155,000,000, the challenge to build 2,000 100 MWe reactors in ten years seems well within our capabilities.

CODA

88. Projections from developers of small nuclear reactors, such as Hyperion, are suggesting that four modular 25 MWe reactors can produce the same amount of energy for $200 million.

APPENDIX

89. John Gilleland, CEO, TerraPower.

90. "One of the stock plots of science fiction was...robots were created and destroyed by their creator. Knowledge has its dangers, yes, but is the response to be a retreat from knowledge? Or is knowledge to be used as itself a barrier to the dangers it brings? With all this in mind I began, in 1940, to write robot stories of my own—but robot stories of a new variety. Never, never, was one of my robots to turn stupidly on his creator for no purpose but to demonstrate, for one more weary time, the crime and punishment of Faust" (Isaac Asimov).

Index

TWR (traveling wave reactor),
173, 201–2

underground containment and
storage, 78, 122–26, 195–96,
197, 198, 200, 204–5
Union of Concerned Scientists
(UCS), 45, 149
United Nations (UN), 54, 64,
103–4, 228, 237, 239, 243
Food and Agricultural
Organization (FAO) of, 37,
99, 275n
uranium, 49, 132, 199, 268
enriched, 155, 159, 173, 201–2
uranium nitride, 196
urban landfills, 25–26, 102
Urey, Harold, 45
USSR, see Soviet Union
Utah, 83

Vietnam War, 52
volcanic events, 22, 125, 213,
270

walking, kinetic energy of, 96
Wall Street Journal, 141
wars, future, 141–42
nuclear, 13, 52, 148–50,
280n–81n
over water access, 169–70
Washington, Warren, 217–18
waste disposal, 39, 40, 103, 130–
31, 279n
toxic, 130, 274n
see also nuclear waste
waste heat, 70–71
water, 10, 29, 32, 40, 65, 124,
125, 141, 241, 270, 282n–83n

desalination of, 36, 168–71
of free-flowing streams, 91
in meat production, 101–2
natural circulation of, 198
natural cycle of, 87, 88
in natural gas fracking, 105–8,
180–81
as nonrenewable resource, 169
nuclear reactors cooled by,
143–44, 155, 157, 166, 198,
201, 222, 240, 282n
rainfall, 31, 217–18
rainwater, 78–79, 87
saline, 85, 107, 169
solar power generation usage
of, 115–16
storage of, 78–79
weight of, 86–87, 89, 91–93
see also hydroelectric power
water fountains, 97
water pollution, 33, 69, 107, 169
geothermal, 85, 86
by meat animal wastes, 103
from natural gas fracking, 105–
8, 181
water rights, battles of, 168–70,
171
water shortages, 69, 100, 168–71
weather-generated, 87–88,
89, 93
Watt, James, 72–73
Wealth of Nations, The (Smith),
277n
Weinberg, Alvin, 44, 49, 50, 124–
25, 153, 172, 181, 190, 205,
272n
Weinberg, Richard, 153
Westgard, Rolf, 120
wet cooling, 115–16